Project AIR FORCE

AIR EDUCATION AND TRAINING COMMAND COST AND CAPACITY SYSTEM

Implications for Organizational and Data Flow Changes

Thomas Manacapilli

Bart Bennett

Lionel Galway

Joshua Weed

Prepared for the
UNITED STATES AIR FORCE

RAND

The research reported here was sponsored by the United States Air Force under Contract F49642-01-C-0003. Further information may be obtained from the Strategic Planning Division, Directorate of Plans, Hq USAF.

Library of Congress Cataloging-in-Publication Data

Air Education and Training Command cost and capacity system : implications for organizational and data flow changes / Thomas Manacapilli ... [et al.].
    p. cm.
    Includes bibliographical references.
    "MR-1797."
    ISBN 0-8330-3503-7 (pbk. : alk. paper)
    1. United States. Air Force. Air Education and Training Command—Evaluation.
2. Aeronautics, Military—Study and teaching—United States—Evaluation. I.
Manacapilli, Thomas.

UG638 .A65 2003
358.4'15'0684—dc22

                                2003024475

The RAND Corporation is a nonprofit research organization providing objective analysis and effective solutions that address the challenges facing the public and private sectors around the world. RAND's publications do not necessarily reflect the opinions of its research clients and sponsors.

RAND® is a registered trademark.

Published 2004 by the RAND Corporation
1700 Main Street, P.O. Box 2138, Santa Monica, CA 90407-2138
1200 South Hayes Street, Arlington, VA 22202-5050
201 North Craig Street, Suite 202, Pittsburgh, PA 15213-1516
RAND URL: http://www.rand.org/
To order RAND documents or to obtain additional information, contact
Distribution Services: Telephone: (310) 451-7002;
Fax: (310) 451-6915; Email: order@rand.org

The goal of this study was to help establish the strategic design for a comprehensive system to assess and manage the cost and capacity of the Air Force's pipeline for enlisted technical training. The study team concluded that such a system is useful only insofar as it supports the decision processes necessary for managing effective training. Therefore, this report examines training management and decision processes to determine the need for data to support informed decisionmaking. It briefly reviews training management systems and associated organizational arrangements in the other services and the private sector to draw insights for a model management system for the Air Force. The study identifies impediments to training planning and management in the current Air Force organizational structure that inhibit the flow of cost and capacity data and hinder effective decisionmaking. It also outlines analytic developments that could help convert raw data into information useful for decisionmakers.

The research reported here was sponsored by the Air Education and Training Command (AETC/CV) and HQ Air Force Deputy Chief of Staff, Personnel (AF/DP) and conducted within the Manpower, Personnel, and Training Program of RAND Project AIR FORCE (PAF) at the RAND Corporation. Earlier, PAF explored the requirements of a technical training schoolhouse model to address pipeline capacity. By explicitly capturing resource limitations and uncertainty, the study team anticipated that this simulation tool could assist AETC in making difficult allocation decisions. In part, the current report confirms the need for such tools and the roles they might play.

This report should be of interest to leaders and staffs concerned with planning and managing the Air Force's technical training pipeline, including AETC organizations, HQ 2AF, and the training wings and squadrons down to the individual course level. The study reflects feedback from this Air Force community. This report covers the time period of October 2001 to September 2002. Please direct questions, suggestions, or other feedback to the lead author at manacapi@ rand.org.

RAND Project AIR FORCE

RAND Project AIR FORCE (PAF), a division of the RAND Corporation, is the U.S. Air Force's federally funded research and development center for studies and analyses. PAF provides the Air Force with independent analyses of policy alternatives affecting the development, employment, combat readiness, and support of current and future aerospace forces. Research is performed in four programs: Aerospace Force Development; Manpower, Personnel, and Training; Resource Management; and Strategy and Doctrine.

Additional information about PAF is available on our Web site at http://rand.org/paf.

CONTENTS

FIGURES

TABLES

The mission statement of the Air Education and Training Command (AETC)—"recruiting, training, and educating professional airmen to sustain the combat capability of America's Air Force"—provides a good starting point for developing information requirements for training management. Combat capability is directly affected by the quantity and quality of trained personnel. And the provision of sufficiently trained Air Force personnel relies upon effective management of training production and, in turn, the cost and capacity of the training system. Arguably, AETC currently has difficulty assembling and using cost and capacity data in managing its training pipeline, particularly for technical training. We find that this is partly due to an organizational structure that is both too complex and too unclear and has overlapping decisionmaking responsibilities.

We developed a four-level model of management to evaluate the flow of data in the AETC training pipeline:

1. The *corporate level* validates and arbitrates training requirements. (See p. 18.)

2. The *strategic training management level* concentrates on the training system's long-term effectiveness. (See p. 19.)

3. The *training management level* handles the day-to-day operations of training. (See p. 20.)

4. The *direct training level* delivers training in the classrooms. (See p. 21.)

Most data needed for informed decisionmaking in AETC exist at the bottom two levels but often do not flow adequately to the top two levels. Part of the problem is that strategic training management is split among multiple organizations: no central organization has the manpower to work capacity issues (e.g., addressing surge or limiting constraints), reduce Trained Personnel Requirements (TPR) short-falls, evaluate quality information, develop cost methodologies for planning, and serve as the single advocate for technical training in the Air Force. As a result, data flow among training management organizations is ad hoc. (See pp. 33–53.)

We looked at how strategic training management was handled in other training organizations to help motivate our model and provide lessons for AETC. The Army has organized strategic training management at the functional level, with no intervening organizations between it and training management. Currently, the Navy has a very decentralized training operation but is conducting an extensive revision effort to correct disconnects discovered during its Executive Review of Naval Training. Our case studies of four major companies with large training programs show that although these companies employ different organizational designs, all have a clearly defined senior person responsible for organizing training and making strategic decisions. (See pp. 14ff.)

We recommend organizational and process changes at the strategic training management level (mostly residing in HQ AETC). We believe that a consolidation of the strategic management functions, within an organization probably headed by a two-star general, would, among other things, resolve many current data flow problems. We also recommend that methodological tools be developed, including simulations to evaluate tradeoffs in the training pipeline, in order to improve data combination and interpretation, particularly in the area of cost. It is also clear that AETC should have a central data "warehouse" for collecting cost and capacity data. We believe that a "real-time" minute-by-minute data tracking system is not warranted and would not be cost-effective. Finally, we recommend that cost and capacity data be fit into the AETC Decision Support System/Technical Training Management System (ADSS/TTMS) architecture already under development for training production data. (See pp. 56–60.)

ACKNOWLEDGMENTS

A number of people contributed to this research. We acknowledge them in a somewhat chronological order.

We began with a schematic of the training process that Al Robbert,[1] a former AF/DP division chief, developed. This schematic was a very helpful guide for our discussions with the offices we visited.

We thank all at HQ AETC who helped us, whom we list here by office: XP (Col Mike Snedeker, Jack Wilder, Meg Salvadore, Jerry Aho, Gary Beitzel, Dan Goetz, Ruth Lease, Capt Bill Crooks, Larry Hughes, Capt Courcy, Richard Carroll, Woody Roberts, Bill Friday, Marybeth Coffer); DO (Robby Robinson, Lt Col Eddie Billman, Tom Gates, Joe Pollock); FM (Pam Struzyk, Cherlyn Koehler, Mark Parsons, Richard Hutchins); CE (Maj Jon D'Andrea, Chuck Miller, Al Ennis, Charles Lasley, Susan Higgins); and RS (MSgt Ed Heinbaugh, Capt James Feldhaus). Our visit to BMT (Lt Col Ada Conlan and the staff of the 319th Classification Squadron) was also of great help.

At the Air Staff we received similar assistance from XP (Capt Guillermo Palos); DP (Col Nancy Weaver, Lt Col Tony Henderson, Lt Col May, Maurice Lang, Col Linda Cunningham, Ken Spires, Capt Joe Cloeter); and AFPOA (Maj Steve Forsythe, Maj Matt Santoni).

Second Air Force, the Keesler AFB training wing, the training group, and the training squadrons were very supportive, and we are grateful

[1]A. Robbert et al., "Determining Course Content for Air Force Enlisted Initial Skill Training: How Can the System Be Made More Responsive?" RAND Corporation, Santa Monica, CA, internal draft, Jul 2001.

for first-class treatment from HQ 2AF (Maj Gen John Regni, D. W. Selman, Sherry Hernandez, Mary Hancock, Col Jim Cropper, Gary Peterson); 81st Training Wing (Brig Gen Mercer, Col J. B. Murray); 81st TRG (Col Fossen, Lt Col Howard Triebold); 81st TRSS (Lt Col Glenda Reichler); 332nd TRS (Lt Col Jackson and his staff); and the 338th TRS (Maj Reese and his staff).

Additionally, we met with retired and former senior personnel with tours in Personnel and in Training. These conversations helped us to understand the evolution in the training organization and process. We thank Jay Tartell, and we thank Gary Teer for providing considerable historical perspective on changes in AETC.

We also thank AETC/HO for tremendous support. The History Office is a gold mine of material.

Also, we appreciate CDR Anthony Cooper and his staff at CNET PAO for their help in understanding the Navy and its recent efforts in the area of training management.

In industry, we wish to thank Shawn Clark, Director of Technical Training, Northwest Airlines; Ed Bales, Motorola University; Keith Hertzenberg, Vice President and General Manager of Training Systems and Services at Boeing Integrated Defensive Systems, Aircraft and Missiles; Keith Jostes, Manager of Technical Training, Boeing; and Sherman Jaffe, Senior Manager of Training Management and Certification.

Our special thanks go to Brig Gen (Ret.) Karen Rankin and RAND colleague Georges Vernez for their thoughtful reviews of the draft report.

Most of all, we thank Lt Col Jim Boyd for his tireless help and support. He set up all our trips and turned around our requests in record time.

ACRONYMS AND INITIALISMS

2AF	Second Air Force
ADO	assistant deputy of operations
ADSS	AETC Decision Support System
AETC	Air Education and Training Command
AETMS	AETC Training Management System
AFMC	Air Force Materiel Command
AFSC	Air Force Specialty Code
AFTMS	Air Force Training Management System
AMU	Automobile Manufacturing University
BMT	basic military training
BOS	base operating support
BST	business support team
BX	base exchange
CBT	computer-based training
CFETP	Career Field Education and Training Plan
CFM	career field manager
CINCPACFLT	Commander in Chief, Pacific Fleet
CNET	Chief of Naval Training
COSMOD	Cost Modeling System

CRE	course resource estimate
DEP	Delayed Entry Program
DoD	Department of Defense
DPG	Defense Planning Guidance
DPL	Deputy Chief of Staff for Personnel, Learning
DPR	Director of Personnel Resources
DSS	decision support system
EIS	Enlisted Initial Skills
ERNT	Executive Review of Naval Training
ERP	enterprise resource planning
FEQ	Field Evaluation Questionnaire
FI	field interview
FY	fiscal year
GAS	Graduate Assessment Survey
GTEP	Guaranteed Training Enlistment Program
HQ	Headquarters
ICW	interactive courseware
IIT	ineffective in training
MAJCOM	major command
MILCON	military construction
MILPDS	Military Personnel Data System
MRT	mission ready technician
MU	Motorola University
NAF	Numbered Air Force
NCO	noncommissioned officer
NW	Northwest Airlines
O&M	Operations and Maintenance

O&S	Operations and Support
OJT	on-the-job training
OMS	Occupational Measurement Squadron
OPNAV	HQ U.S. Navy Staff
OPTEMPO	operating tempo
OSD	Office of the Secretary of Defense
PGL	Program Guidance Letter
POI	plans of instruction
POL	petroleum, oil, lubricants
POM	Program Objective Memorandum
QOL	quality of life
ROI	return on investment
SAT	student awaiting training
SC&FG	Signal Center and Fort Gordon
SME	subject matter expert
SOT	student out of training
Stan/Eval	Standards/Evaluation
STS	Specialty Training Standard
TCO	total cost of ownership
TD	training device
TDS	Training Decisions System
TIDES	Training Impact Decision System
TPR	Trained Personnel Requirements or Training Program Requirements
TPS	Training Planning System
TRADOC	Training and Doctrine Command
TRG	training group

TRS   training squadron

TRSS   training support squadron

TRW   training wing

TT   technical training

TTMS   Technical Training Management System

U&TW   Utilization and Training Workshop

Chapter One

INTRODUCTION

The purpose of this study was to help improve the responsiveness of the U.S. Air Force's technical training system to changing force and support requirements. Initially, we focused on developing an understanding of costs and capacity constraints within the training system with an eye toward providing better support for strategic decisionmaking. We quickly discovered, however, that various impediments stood in the way of accumulating this information. Almost always, the raw data required to support decisionmaking existed somewhere, but the task of collecting them and processing them into information useful to a decisionmaker was formidable. Obstacles to resolving these issues include the quantity of data, the diverse ad hoc databases maintained at various organizational levels, methodological challenges for turning raw data into useful information, and excessive layers of organizational management that inhibit the flow of data. Although efforts to develop on-line systems have recently been able to reduce some of these problems, many still remain. In this report, we provide a variety of recommendations for how to enhance technical training management and planning by improving the availability and accessibility of cost, capacity, and other information needed to support decisionmaking.

Air Education and Training Command (AETC) is unique among major commands (MAJCOMs) in its mission and role. Unlike most MAJCOMs, AETC focuses on preparing for warfighting rather than on warfighting itself. Perhaps surprisingly, AETC has the second largest aircraft fleet among the commands and the smallest budget. It also has the largest number of people passing through its gates (considering both transient and permanent-party personnel). Estimating the cost of this command's operation is not a simple process. The numbers of students continually change, the curriculum can change, the costs of supplies change, the numbers of instructors needed vary, and so on. AETC has the primary responsibility for evaluating training policy decisions, particularly the cost of training. To do so, it needs a reasonable estimate of cost per graduate based on varying conditions.

AETC also must develop and maintain a responsive training pipeline. National security sometimes demands a rapid buildup of forces. Conversely, changes in technology sometimes reduce manpower requirements. The training pipeline must be responsive to manpower changes that can come about in many ways: acquisition of new aircraft or space systems, merger of career fields, introduction of new technology, privatization or outsourcing of certain skills, etc. This dynamic environment makes planning and programming very difficult.

INFORMATION AND STRATEGIC MANAGEMENT OF AIR FORCE TRAINING

Like many commercial companies, the Air Force is a complex organization that uses a myriad of skills to accomplish its mission. Such organizations need a comprehensive program of initial training to give new employees the skills needed for their jobs, plus continuing training to help current employees upgrade their skills for promotion and keep up with new techniques. For many companies, the importance of systematic, comprehensive training has been a recent discovery, resulting in the establishment during the 1990s of "corporate universities" to manage and provide employee education. In contrast, the importance of training has long been appreciated in the Air Force, whose extensive training establishment develops personnel from their basic training at entry to the Air Force, through initial skills training, to advanced technical and management education. Air Force training is a vital part of the "blueing" process that inculcates the Air Force culture into its people.

Like most other organizations with a major training component, the Air Force has traditionally focused on the provision of training (e.g., classrooms, instructors) and its direct support (e.g., building maintenance, dormitories, dining halls). So-called strategic management traditionally involved assembling the annual or biennial training budget. But rapid security-environment changes that began with the end of the Cold War and included major force realignments and the introduction of new technologies (e.g., computers, ubiquitous digital electronics, stealth) created a need for more-active strategic management. This includes the advent of new courses, the introduction of new Air Force Specialty Codes (AFSCs),[1] and the retirement of old ones. At the same time, the military drawdown led to diminishing budgets, stricter justifications for the training budget, and a heightened interest in efficiency. These changes increased the need for an expanded strategic management function that could weigh alternatives, estimate costs and capacities, and effectively allocate training resources across all Air Force training elements. This new emphasis requires a fresh examination of the organizations that currently have strategic management responsibilities and of the information they need to discharge their responsibilities.

IMPLEMENTING ENHANCED STRATEGIC TRAINING MANAGEMENT

There are a number of major stakeholders in the Air Force's strategic training management. One responsibility of the office of the Deputy Chief of Staff for Personnel (AF/DP) is to estimate training requirements based on current and projected force structure needs, taking into account personnel policies, such as promotions and special pays, that are also largely under its control. AF/DP is the overall advocate for training within the Air Staff, which is the corporate Air Force management organization.

Across the Air Staff, in many different offices, are the career field managers (CFMs). Their goal is to ensure that the training requirements are appropriate for the operational tasks required, and that the quality of the training is sufficient for the operational tasks required.

Additionally, the office of the Deputy Chief of Staff for Air and Space Operations is concerned that the mix and quantity of trained airmen are sufficient for the operational requirements.

---

[1]AFSCs are the codes assigned to various occupations within the Air Force.

AETC provides most of the training in the Air Force. It runs the class-rooms, trains the instructors, plans and modifies the curriculum, and is responsible for executing the training process within the budget. Because of this role, AETC is the primary collector and developer of most of the data needed to make training management decisions, including data on costs, capacity, production quantity, and training quality. AETC is also responsible for passing suitably aggregated data to the Air Staff for AF/DP's use.

Unfortunately, both AETC and AF/DP have been dissatisfied with the timeliness, accuracy, and accessibility of these data. In some cases, data needed to produce information desirable for decisionmaking are not collected at any level (are unavailable), or the methodology for generating such information from raw data is insufficient or un-known. Aggravating this problem is the fact that the means and methods for capturing data at its lowest level are often inconsistent. As a result, there are many local independent databases for specific classes or at specific schools, most of which are computerized, but some of which exist only on paper.

To standardize much of the data collection, AETC is building the AETC Training Management System (AETMS) to track different aspects of AETC instruction. For example, the Technical Training Management System (TTMS) within AETMS records student course performance and status for all technical trainees, curriculum plans, instructor data, and information on training devices. The AETC De-cision Support System (ADSS) then adds the capability to "roll up" these data for use in making training management decisions. Unfor-tunately, TTMS is still being populated, so its use as a warehousing mechanism and ADSS's use as a processing tool are yet to evolve.

These developments represent significant progress toward resolving some of AETC's most challenging data problems. However, even when they are fully implemented, information on such key items as capital costs and facility capacity will still be lacking. Thus, aggregate cost, capacity, quantity, and quality data for responsive strategic decisionmaking will remain elusive. Decisionmakers have difficulty getting answers that would help them make better-informed deci-sions. For example, how many more trainees in one AFSC—say, se-curity forces—can be brought in before classroom and infrastructure requirements at the training base are exceeded? Furthermore, the total cost to train, including the fixed costs of facilities and training devices, cannot be precisely determined. Deriving this kind of essential information for strategic decisionmaking is yet another step beyond the significant data management systems currently being implemented.

PROJECT EVOLUTION, METHODOLOGY, AND LIMITATIONS

This project was initially undertaken to assess the needs for cost and capacity data in order to support strategic training management at AETC and the Air Staff. One of our first steps was to focus solely on technical training rather than on both technical and pilot training. Technical training is by far AETC's largest workload, with approximately 35,000 initial skills students and 1,200 courses annually. Pilot training, although a larger share of AETC's budget, is much simpler in structure. Furthermore, pilot training has received more attention in the past, so there is already a significant set of tools for managing cost and capacity. Thus, the technical training system constitutes the greater need.

To evaluate the need for cost and capacity data at the strategic management level, we began by conducting a series of interviews at HQ AETC (XP, DO, FM, CE, and RS), the Air Staff (XP, DP, and AFPOA), Basic Military Training (BMT) (the 319th Classification Squadron), and Second Air Force (2AF). To represent the process and concerns at lower organizational levels, we interviewed training leadership and staff at Keesler Air Force Base: the 81st Training Wing (TRW), the 81st Training Group (TRG), the 81st Training Support Squadron (TRSS), and the 332nd and 338th Training Squadrons (TRSs). Our findings from these discussions are tempered because of the diversity among training suborganizations. Additionally, we met with retired and former senior personnel with tours in Personnel and in Training to better understand the organization's evolution and its process. All of these conversations, although naturally reflecting the viewpoint and biases of those interviewed, provided us with a great number of insights into the current capabilities and limitations of the development, processing, and propagation of data within AETC.

Although the interviews were somewhat informal, we asked basic questions, such as: What strategic training decisions need to be made and by whom? How could more accurate cost and capacity data be used to support these strategic decisions? We interviewed key people in a wide range of offices supporting AETC's strategic management process and reviewed associated AETC strategic management policies.[2] It became clear that cost and capacity data must also be linked with student enrollment and production (quantity) and training assessment (quality) data already being recorded in TTMS, since all four types—cost, capacity, quantity, and quality—are needed for most strategic management decisions. We also found that much of the raw cost and capacity data were in fact collected and available at lower levels of AETC management. However, access to these data by the higher levels was sometimes blocked by the size and cumbersome structure of the data and the many ad hoc systems used to capture them.

In addition to these impediments, we found two other problems:

1. Methodological limitations. To make use of fairly detailed data for decisions at the strategic level, new analysis tools and, in some cases, methodological innovations are required.

2. Organizational impediments. The organizations responsible for management of Air Force training are fragmented, which reduces their ability to share data.

The problem of methodological limitations may be resolved by developing a more comprehensive set of tools in a decision support system (DSS).[3] The problem of organizational impediments is more difficult to resolve and will require looking at "the big picture" of AETC technical training.

We also reviewed the training literature and interviewed leaders from other training organizations to better understand how different organizations have dealt with these issues. Those reviewed include the Army's Training and Doctrine Command (TRADOC), the Navy's

---

[2]For example, HQ USAF/DPDT, AFI 36-2201, Vols 1–6, Sep–Oct 2002.

[3]See, for example, Ralph H. Sprague, Jr., and Eric D. Carlson, *Building Effective Decision Support Systems*, Prentice-Hall, Englewood Cliffs, NJ, 1982.

training command (Chief of Naval Education and Training, or CNET), and training units inside large industrial organizations—Motorola, Northwest Airlines, Boeing, and Automobile Manufacturing University (AMU). This report consolidates comments from these documents and discussions and provides support for our findings and recommendations.

To provide a more holistic approach to resolving these issues, we first step back to take a broad look at AETC's role and mission and the context of strategic training decisionmaking.

VISION FOR EFFECTIVE STRATEGIC TRAINING MANAGEMENT

According to AETC's mission statement, the training management system is responsible for "recruiting, training, and educating professional airmen to sustain the combat capability of America's Air Force."[4] We take this mission statement to be the guiding principle in defining the vision for a system (including organizational structure and information management responsibilities) designed to better inform AETC decisionmakers.[5] We particularly focus on the desired result: "to sustain the combat capability of America's Air Force."

Applying the Vision

To understand how this vision applies, we first take a very broad, simple look at combat capability. Figure 1.1 depicts two main ingredients of readiness that sustain combat capability: ready personnel and ready equipment. Training quality and quantity are impor-

---

[4]From AETC Strategic Plan, 2001, and updated on the AETC Web site at http://www.aetc.randolph.af.mil.

[5]AETC's initial effort to improve cost and capacity information supporting strategic decisionmaking, as explained in Appendix A, emphasized "agile" and "integrated." An agile decision support system gives commanders the ability to respond in a timely fashion to internal organizational and external world events. An integrated decision support system looks at the whole training pipeline, from the determination of recruiting requirements to the absorption and on-the-job training (OJT) of airmen in their first operational units. Agility and integration are important attributes, but we believe they are only part of the vision for a comprehensive cost and capacity system in that they do not define a result.

RAND *MR1797-1.1*

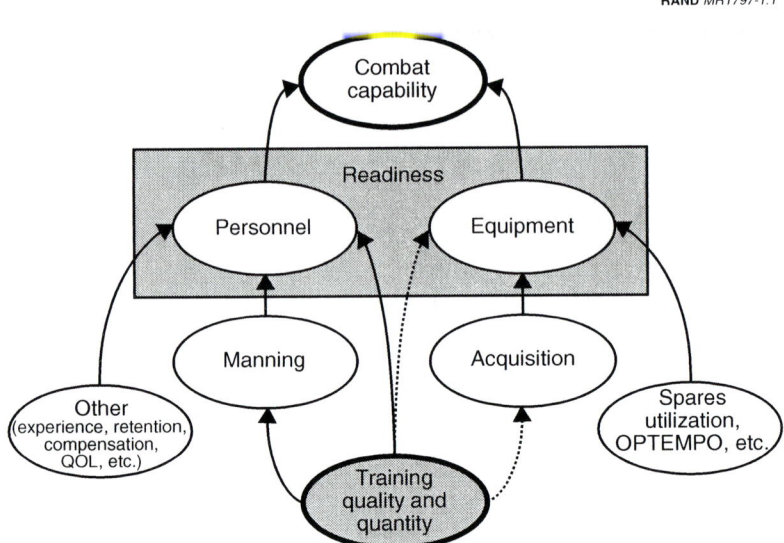

**Figure 1.1—Sustaining Combat Capability with Ready
Personnel and Ready Equipment**

tant determinants for both. Although other factors, such as quality of
life (QOL), also influence personnel readiness, training quality has a
direct impact. Additionally, training quantity is essential for meeting
manning requirements and hence is a significant part of personnel
readiness. Training quantity and quality also impact equipment
readiness, although somewhat less directly. High-quality trained
personnel are more capable of rapidly acquiring and more thoroughly maintaining the equipment. Thus, AETC's responsibility for
the quality and quantity of airman training is central to enhancing
and sustaining combat capability.

Because of training's direct relationship to personnel readiness, we
next examine this component in more detail. Figure 1.2 shows this
relationship and how quality, quantity, cost, and capacity are linked.
As highlighted in this figure, the resources spent on increasing
training quality or quantity could also be spent on compensation
initiatives, retaining experienced personnel, expanding the number
of weapon system support items, or other factors that would improve
readiness. Resource trades of this nature need to be done at the

RAND *MR1797-1.2*

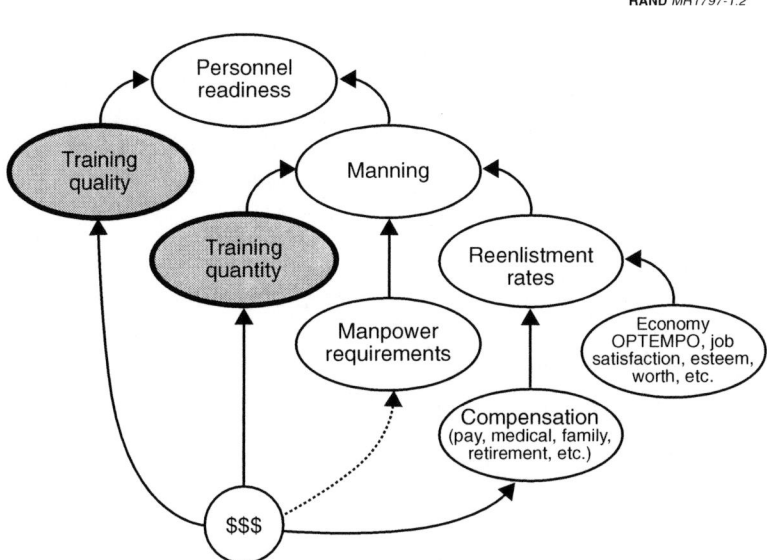

Figure 1.2—Training-Related Tradeoffs in Sustaining Combat Capability

highest levels of Air Force management—what we call the *corporate level*. To fully understand the implications of resource decisions at the corporate level, information on each option is needed for tradeoff comparison. As these figures suggest, the tradeoffs between training and retention are linked to complex interactions within the training and manpower systems and depend upon cost, capacity, quality, and quantity data. If organizations with training oversight at the corporate level are too fragmented (multiple offices with poor linkage), information to make efficient and effective resource trades will be significantly impeded. As a result, the Air Force may not achieve the most cost-effective or desired outcomes. A corporate-level focal point for decision support data would better enable informed corporate tradeoff decisions.

Relationship of Trained Personnel Requirements to the Vision

The quantity of specially trained airmen needed to sustain high readiness and replenish the Air Force's combat capability is determined through the Trained Personnel Requirements (TPR) process. The corporate Air Force combines the requirements (current and projected) of all the users, MAJCOMs, Guard, reserve, and non–Air Force users (other services, joint organizations, and nations) and uses current retention rates, manning, projected cross flows, and prior-service gains to produce a requirement listing by individual AFSC. This initial TPR is the estimated "real" requirement to sustain the combat capability of the Air Force.

But initial TPR numbers are not limited by AETC's cost or capacity constraints. Prior to the Training Flow Management Conference (more commonly referred to as the TPR conference), the Air Staff (AF/DPLT) sends an Enlisted Initial Skills (EIS) Program Guidance Letter (PGL) to the training schools (squadrons within the AETC organization). The schools reply with a list of constraints by course and options for resolving the constraints where possible. AF/DPLT then issues a revised EIS PGL, also prior to the Training Flow Management Conference. These and other restrictions are considered at the conference, after which a "final" TPR is developed. While the PGL that is the final output of this conference serves as the official tasking document, it actually reflects the constrained requirement for trained personnel rather than the true, "real" requirement. In reality, the final TPR is only that portion of the real TPR (the true requirement for sustaining combat capability) believed to be feasible. Compounding this problem is the fact that without accurate and sufficient cost or capacity information, the constraints put on the "real" TPR may be inaccurate, and thus may cause difficulties in the actual execution of the TPR. More importantly, the final TPR may be excessively constraining and therefore not able to completely sustain the combat capability of the Air Force as necessary. Accurate and appropriate cost and capacity data would help the Air Force identify and reduce constraints on production and hence on the ability to replenish combat capability. This is where our work examining AETC's cost and capacity system ties in with improving AETC's strategic decisionmaking in support of its mission statement.

REPORT STRUCTURE

In Chapter Two, we begin by describing an idealized training management system based on our examination of Air Force and other military and corporate organizations. Within this structure, we describe where data should be collected and how they should be disseminated. This leads to an examination of the current AETC organizational structure in Chapter Three. We compare the idealized structure to the current AETC organizational structure and comment on differences. We also examine the data flow among and between management levels and note deficiencies in the availability and accessibility of information necessary for decisionmaking at each level. We offer recommendations for improving the information systems behind the decisionmaking process.

In Chapter Four, we summarize our research, analysis, and recommendations, and we outline additional work to support AETC decisionmaking. In particular, we highlight the beginning of a modeling suite to examine alternative policies and practices either at the schoolhouse level or as part of the end-to-end training management system. Two appendices provide detailed information on various aspects of this study, including detailed insights from other organizations that wrestle with similar training issues.

Chapter Two

MODEL TRAINING MANAGEMENT STRUCTURE AND ASSOCIATED INFORMATION NEEDS AND FLOWS

As discussed in Chapter One, the initial purpose of this study was to help improve the responsiveness of the Air Force's technical training system to changing force and support requirements. More specifically, the goal was to enhance the flow of cost and capacity data to help ensure that management decisions would be made in an informed manner. Accordingly, the strategic vision presented in Chapter One requires careful planning based on having the right information from the right data at the right time. This means that behind the data flow is a management structure that provides contact points for source data processing and aggregation, and receiving decision points for the information. So, before addressing the specific data flow paths, it is worthwhile to step back and ask what the organizational structure of training management should be like. With the organizational structure in place, we can more easily determine what information is needed at the different levels within the structure and where the data to provide that information can or should originate from.

A fundamental question for the management structure is how many levels of organization are necessary and what should their functions be? For example, the training management that occurs at HQ AETC is very different from the training management that occurs at the classroom level. These two distinct management levels are concerned with different types of decisions, timescales, and types and aggregations of data useful in making informed decisions. Although the classroom is the primary source of the data needed for most de-cisionmaking, these data need to be manipulated and processed into useful information for decisionmakers at other levels.

The handling and filtering of data from the classroom up through higher management levels can be inhibited by organizational struc-ture, especially unclear or overlapping institutional responsibilities. Therefore, in thinking about enhancing the flow of cost, capacity, and other data, it makes sense to consider an exemplary manage-ment system. For this exemplar, we define the unique functions of different levels and the data needs for these levels with as little refer-ence as possible to the current Air Force management structure. We motivate our exemplar by first examining the training management structure of other organizations: those of other services and a set of analogous commercial organizations. We then discuss the deci-sion problems associated with the different management levels and, finally, the data flows to support these levels.

TRAINING IN THE ARMY, NAVY, AND SELECTED COMMERCIAL FIRMS

We interviewed and/or visited U.S. Army and Navy training man-agement organizations as well as a selected set of commercial train-ing organizations: Automobile Manufacturing University (a pseudonym for the training arm of a major automobile manufac-turer), Motorola University, Northwest Airlines Technical Training, and Boeing Integrated Defensive Systems Technical Training. The commercial training organizations are substantially smaller than their military counterparts. Additionally, their focus is not predomi-nantly on entry-level skills, which largely drive the physical atten-dance requirements in the service organizations. However, their strategic management structure does have relevance to the Air Force.

Details of our findings are in Appendix A in the section on case stud-ies. Here we provide an overview of the details most relevant to the training management structure. The overview is presented as four lessons learned.

Lesson 1: Consolidate Management Functions

The Army provides an example of the variety of benefits that can ac-crue from a decentralized, consolidated organizational structure. Because of its distinctive division into warfighting functional areas, the Army combines its training and doctrine development into relatively autonomous "schoolhouses" for each of these areas. All schoolhouses are commanded by senior Army officers and report to the TRADOC, which is under a four-star general. TRADOC provides general training policy and budget coordination, but responsibility for training in the functional area is almost entirely given to the schoolhouse. Consolidation of management functions within the schoolhouse leads to efficiency in decisionmaking within the functional areas and ease in identifying and responding to data needs. On the downside, the decentralization and consolidation often cause difficulties at the next higher level, where integration across functional areas must be performed. This suggests a need to consolidate management functions so as to balance cross-functional training needs and the operational management of the schoolhouse.

In industry, the organizational structures differ from each other, but there is a clearly designated senior person responsible for training strategy and execution. Administrative staffs for training manage-ment seem to be very lean, at least partly because functional respon-sibility is clear.[1] This is even true of commercial training structures that are *more* decentralized than those of the Air Force. The single focal point for strategic training management within industry high-lights the value of not splitting strategic management among organi-zational levels.

Lesson 2: Reduce Organizational Layering

The Army's structure indicates that a reduced amount of organiza-tional layering is beneficial for improving many decisionmaking functions, although such an arrangement is prone to stovepiping. The Navy's decentralized structure is taking the blame for Navy problems in training and suggests that over-decentralization can

---

[1]We did not determine whether some training management functions were performed in other parts of these organizations.

lead to problems. The Navy is currently reorganizing its training because of its dissatisfaction with what it sees as far too much fragmentation.

Although the Navy's situation is somewhat of an extreme example, it nonetheless exemplifies the problems that occur with an excessive management structure. In its old structure, 63 organizations imposed training requirements, 38 performed training management functions, and 39 coordinated training exercises. The Navy found that the lack of coordination in imposing training requirements was particularly costly. It is notable that Northwest Airlines Technical Training, which, like the Air Force, manages training as a corporate function, focuses much attention on making sure its senior management maintains close contact with its customers' evaluations of training. This is facilitated by a flat organizational structure with as few organizational units and decision levels as possible.

Lesson 3: Derive Training Requirements from the User Community

Like the Air Force, all four of the corporations we interviewed get much of their instructional direction from their operational communities. However, the Air Force's operational representatives are the career field managers (CFMs), for whom curriculum assessment is one responsibility among many. Further, in the Air Force's technical training areas, each of these persons is, at most, a senior noncommissioned officer (NCO). In contrast, the responsible person in all the industrial organizations we visited is a senior staff member with considerable formal influence in the organization.

Lesson 4: Span of Control Should Not Be a Big Concern

The case studies suggest that other training organizations are not as concerned as AETC is with limiting span of control. Some of those interviewed gave the need to limit span of control as one reason for AETC's heavy hierarchical structure, with its involvement of several HQ USAF offices, HQ AETC, numbered Air Forces, TRW commanders, and TRG and TRS commanders. Although not a training organization, the Air Force Materiel Command (AFMC) is a large Air Force MAJCOM with an organizational structure much flatter than that of AETC and a much broader span of control for its general officer.[2]

MODEL MANAGEMENT STRUCTURE FOR TRAINING

The lessons learned from operational training organizations conform to the theoretical literature on organizational design and management structuring. This literature identifies three basic layers of management—the strategic apex, the middle line, and the operating core[3]—with support staff and infrastructure acting as buffers. The purpose of this structure is to enable decisionmaking, innovation, and information acquisition and distribution.[4] The fact that two of these—decisionmaking and information acquisition and distribution—are concerns expressed by AETC indicates that a structural examination of training management is in order.

The four lessons learned from our case studies and the theoretical literature led us to the design of a simple exemplar for training management. This suggests that it is useful to think about four levels of responsibility within a training organization:[5] (1) corporate, (2) strategic training management, (3) training management, and (4) direct training. Figure 2.1 illustrates these four management levels and includes as additional key components both the outside users of the training system and the training infrastructure surrounding and supporting direct training and training management. Users are the consumers of training: they generate the requirement for trained personnel. Training infrastructure provides an essential support function for day-to-day training management and direct training but does not constitute a separate management level.

---

[2]As explained in Appendix B, AFMC's structure is an example of the feasibility of delayering and should be examined in more detail.

[3]Henry Mintzberg, *The Structuring of Organizations*, Prentice-Hall, Englewood Cliffs, NJ, 1979.

[4]George P. Huber, "The Nature and Design of Post-Industrial Organizations," *Management Science*, 30(8), 1984.

[5]We divided our strategic apex into two separate entities: corporate and strategic training management. Corporate brokers the various user requirements and balances training against the operational expenses. Strategic training management is our true strategic apex for training.

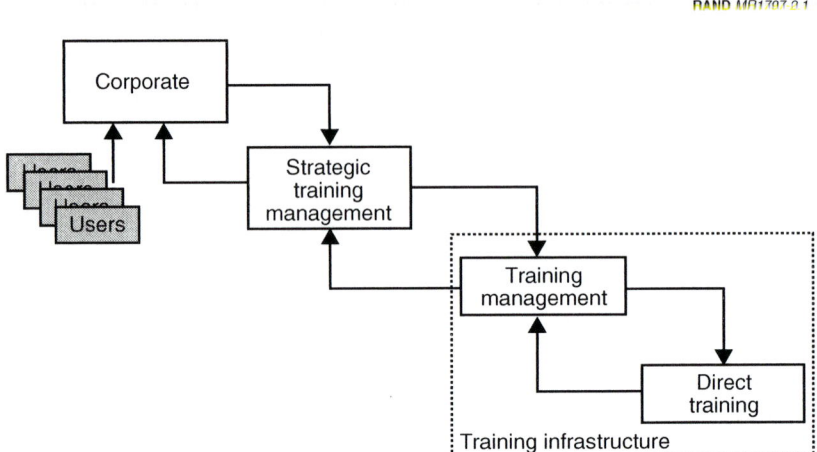

Figure 2.1—Four Broad Functional Levels in the Training Process

Each broad level has a specific span of control and specific functions and data requirements to support its training management decisions. Table 2.1 lists the macro-level functions of each of the levels. We next provide some examples of questions relevant to each level whose answers require decision support data.

Corporate Level

The corporate level determines, validates, and arbitrates all of the users' training requirements. These requirements form the basis for the two fundamental components of the training system: course content and number of trainees needed. The corporate level also manages the overall training budget and sets overall training management policy for the service. It assesses resource tradeoffs among training (quality and quantity) and other resource uses, such as force structure, research and development, quality of life, and manning and compensation policies.

Examples of management decision questions asked at the corporate level are

- What is the right production number for each individual training program?

Table 2.1

Technical Training Functions for Each Training Management Level

Four Main Levels		Training-Related Levels	
Corporate	Determine training rqmts Overall training budget General and AF training policy Validate training rqmt	Users	Training needs

Strategic
training
management
	Training policy
	Standards/evaluation
	Resource allocation
	   Priorities
	   Manpower factors
	Advocacy		

Training management	Resource mgmt Academic faculty allocation Training administration Student services Technology	Training infra- structure	Base operating support Transportation Dorms Dining hall
Direct training	Course development Direct instruction		

- Should we spend our next dollar on recruiting and training new people or on retaining those we already have?

- Are requirements for trained personnel from major users of the organization being met?

Strategic Training Management Level

This level concentrates on the training system's long-term effectiveness. It drafts the entire training budget, sets training policy, allocates resources, and formulates plans for future responsive, effective, and quality training. This level is the primary advocate for improved training.

Examples of management decision questions asked at this level are

- What is the cost to train a person in a specialty area?

- What are the resource constraints limiting the number of people that can be trained?

- Are the needs for training recognized at the corporate level?

- How are future needs evolving?

- What are the long-term impacts of increasing or decreasing throughput?

- What is the tradeoff between capacity and flexibility as training needs change?

- How can training resources most effectively be used to meet and stabilize production quotas?

- Should instructor tour lengths be increased?

- What long-term investments (e.g., facilities, technology insertion) need to be made?

- Is training quality meeting the needs of the users? How does quality impact the cost to train and the amount of time it takes (including washbacks and washouts) to train a needed cohort?

Training Management Level

The training management level handles common functions such as registrar, student services, and basic schoolhouse administration. The focus is on day-to-day operations of training. It manages the training infrastructure's shareable resources (dorms, dining halls, and academic facilities) and apportions them among the direct training units. Training management interfaces directly with the infrastructure for support (transportation, housing, feeding). Finally, such common functions as training device maintenance and technology infusion are located at this level.

Examples of management decision questions asked at this level are

- How should facilities be allocated to each training unit?

- How much and how quickly could a particular course's throughput be increased or decreased?

- Can technology improvements increase the quality of training or decrease the time to train? Are such improvements cost-effective?

- How should training devices be maintained?

- Can economies of scale or new technologies be used across courses to improve the quality or reduce the cost of training?

- Should remedial classes in some areas be considered?

Direct Training Level

This level involves the actual training in the classrooms, covering such functions as class schedule development, instructor placement, handling of students, and development and maintenance of training devices. Basic quality functions (student testing and instructor evaluation) are also performed at this level. This is where real production occurs and where it is measured and annotated in student records.

Examples of management decision questions asked at this level are

- How often can classes be taught?

- Are there enough instructors to teach the classes?

- Are new or improved training methods or devices required to meet new training needs?

- Should students proceed into the next instruction block?

- How should ineffective students be handled?

Users and Training Infrastructure

Training organizations exist to support users who demand trained personnel. Users direct their needs (quantity) and requirements (quality) to the corporate level; they set the demand for training.

We define *training infrastructure* as the organization structure that owns and maintains the shareable resources at a particular installation for all who use those resources, including training. This infrastructure encompasses a wide variety of resources, including roads, security, the hospital, base services, housing, base exchange (BX), the commissary, and recreation. This is exclusively a support function provided to the training management and direct training levels.

Training infrastructure manages the resources and provides them to training management for apportionment. It is independent of any specific management level and is not necessarily a chain-of-command organization. It could report to the training management level, as is done in the Army. Although span of control may be a concern, no additional level of management purely for making infrastructure decisions is needed between strategic training management and day-to-day training management.

In the next section we apply this exemplary framework to AETC in order to better understand what information flows are needed between different levels to support decisionmaking and to organize and clarify the roles of the organizational levels.

INFORMATION NEEDS AND FLOWS BY TYPE AND LEVEL

Chapter One proposes a vision for a cost and capacity system that mirrors AETC's mission statement: sustaining the combat capability of the Air Force. In analyzing this vision, we posited that quantity and quality are its key metrics and that cost and capacity are the tools for management and planning. We now bring together the four broad organizational levels introduced earlier in this chapter with these four main types of data—cost, capacity, quantity, and quality.

We start by looking at the types of questions and issues that are of interest to each management level. We then track the necessary[6] flow of data to support these decisions. While our focus is on an idealized organizational structure and corresponding data flows, the discussion at times refers to flows as they currently exist.[7]

After the "ideal" flow of data is described, the next section maps the parts of the current AETC training management structure to our exemplary organizational levels and evaluates the current data flow to identify deficiencies in data *availability* (existence of the data) and data *accessibility* (ability to get at the data) at each organizational level. This lets us see whether the data flow is obstructed by inconsistencies in the organizational process.

---

[6]*Necessary* is not intended to mean "perfect" or "optimal." Given the structure proposed and the analysis of needed information, *necessary* refers to the data that need to flow to accomplish all the objectives.

[7]Indeed, we should expect some of the data to flow in an idealized manner. AETC has been in the business of training for quite some time.

AETC Management Information Needs and Flows

For training management to make well-informed decisions, various types of data need to be collected and processed. As noted, we categorize data into four broad areas: cost, capacity, quantity, and quality. Most strategic and direct training decisions require combinations of these data items to appropriately choose a course of action. For example, to determine whether the throughput of security forces can be expanded, a decisionmaker would need

- Cost data: instructor and student pay and allowances, classroom support (equipment, devices, and products), and facilities

- Capacity data: classroom space, dorms, and dining halls

- Quantity data: number of students currently in the pipeline, washback/washout rates, and BMT throughput

- Quality data: student skill mastery based on class size and/or course duration.[8]

The following subsections describe each data category in more detail, explain typical decision questions associated with the category, and identify the data sources.

Cost Information Data Needs and Flows

We begin our examination of cost data by describing the types of decisions made at the four levels of training management that need to be informed with cost data.

Training management and direct training deal with the short-term and day-to-day needs of the training process. Decisions for which cost data are needed typically focus on these issues:

- Allocating training facilities and dorms

---

[8]A suggested issue associated with these data is how to pay for the necessary changes. The cost and capacity data can be used to determine the extent of the increase and whether the corporate level or end users should provide the additional resources. The question of how to pay for the increase is a judgment factor and outside the scope of our study.

- increasing or decreasing resources, such as instructors or training devices

- Determining whether operations and maintenance (O&M) for specific systems should be done by TRSS or contracted out

- Supporting technology and innovation in the classroom that lowers overall costs or increases teaching effectiveness

- Enhancing the efficiency of reporting processes and improving individual teacher skills and techniques.

Strategic training management is responsible for setting priorities for capital expenditures and the recapitalization of large-cost items, which include training facilities, base infrastructure, and high-cost training devices. Although the budget for some of these items (such as new dormitories) may not be under AETC's control, strategic training management has a role in determining the need and expressing a prioritization for such projects. This level also oversees systems (and their costs) that support training across bases and AFSCs—e.g., training-the-trainer courses and information and management systems. In addition, strategic training management has a key role in forming decisions on the cost impacts of changing course content or length. Finally, strategic training management needs to know the cost of training for each course or set of courses so it knows how to price them when external users (such as other services, DoD agencies, and foreign nationals) attend the training.

Figure 2.2 depicts the cost data flow. Operations and Support (O&S) costs percolate up from the direct training and training management levels. The most obvious of these costs are associated with direct instruction and course development and include such things as classroom consumables.

The training management level manages funds for O&S, including technology insertion through such capabilities as interactive courseware (ICW) and computer-based training (CBT). It also maintains and repairs training devices for the direct training units. These costs are categorized by O&M costs. Holding these costs at this level allows economies of scale to be exploited across training units in the same geographic location. The training management level also

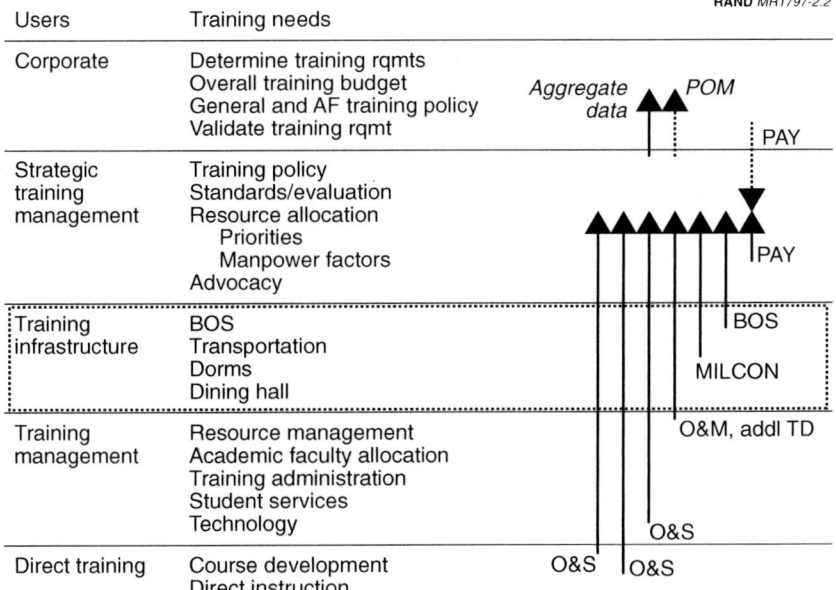

Figure 2.2—Cost Data Flow

manages the resources for building additional training devices (TDs), such as simulation devices or wood replicas.

Training infrastructure manages the costs associated with base contract support for base operating support (BOS) and military construction (MILCON) projects. Training infrastructure also manages and should account for utility costs.

All of these costs should flow up to the strategic training management level, where they form the data for building and justifying the training budget. Strategic training management prioritizes unfunded and new training requirements while honoring the resource conservation principles inherent in technical training.

The budget request then goes to the corporate level. The aggregate supporting data should travel with the budget request to help defend the cost estimates. In the government, pay costs are computed at the strategic training management and corporate levels.

Capacity Information Data Needs and Flows

Capacity decisions primarily address the ability of the individual training bases to instruct, house, and feed all of their students. Capacity decisions must take into account the differing training requirements for AFSCs educated at the base, unit integrity issues, and gender separation requirements. Defining simple measures of capacity can be quite difficult because of the variety of components that make up capacity (instructors, facilities, training devices), the timeliness of these components (e.g., a firing range is only needed for a portion of a course), and the frequent sharing of some major resources among courses (particularly facilities). Furthermore, systems reaching high capacity (utilization rates) often break down very rapidly.[9]

Capacity-oriented decisions made at the strategic training management level require aggregate data from lower management levels at the same geographic location. While capacity problems are manifest and often handled locally, strategic management is still involved in monitoring the issues, advocating improvement, and allocating resources. A decision to add dorm space, for example, should be made at the strategic level because of the substantial cost and potential effect on multiple training management units.

Capacity issues fall into six broad categories: class size, number of instructors, classroom space, training devices, dormitories, and dining facilities.[10] Figure 2.3 depicts the expected flow of capacity data.

The direct training level has readily available data about class size, the number of training devices, and the available instructors. The training management level has a better understanding of the number of classrooms, since it can redistribute some of the generic resources to support production changes in the direct training units. Training infrastructure supports training management by providing data on the capacity of the dining halls and dorm rooms. Additional BOS-related capacity items (such as hospital size, commissary, and BX)

---

[9]It is well known, for example, that queuing systems reaching greater than 80 percent utilization experience a rapid rise in wait times and queue length.

[10]Inherent in these categories are the usage rates of the various resources.

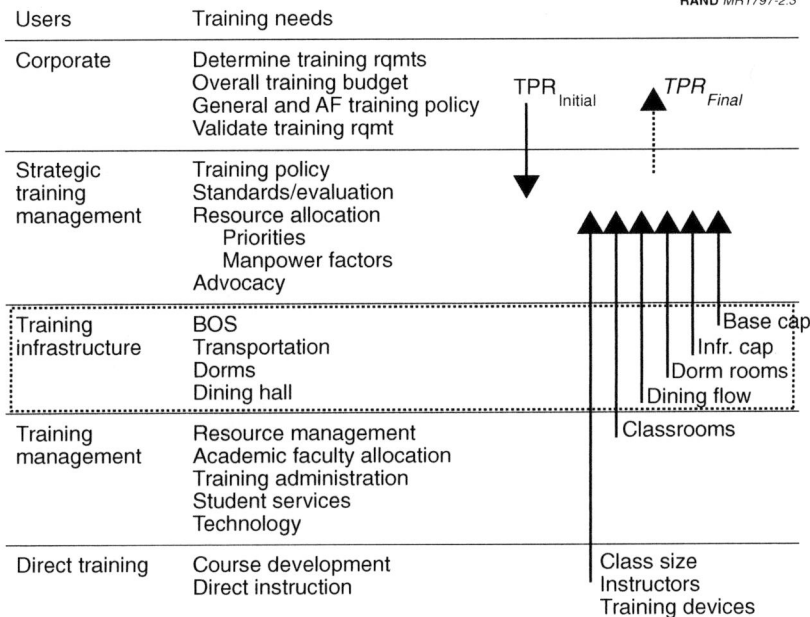

RAND *MR1797-2.3*

Users	Training needs
Corporate	Determine training rqmts Overall training budget General and AF training policy Validate training rqmt

Strategic
training
management
	Training policy
	Standards/evaluation
	Resource allocation
	  Priorities
	  Manpower factors
	Advocacy

Training infrastructure	BOS Transportation Dorms Dining hall
Training management	Resource management Academic faculty allocation Training administration Student services Technology
Direct training	Course development Direct instruction

$TPR_{Initial}$ TPR_{Final}

Base cap
Infr. cap
Dorm rooms
Dining flow
Classrooms

Class size
Instructors
Training devices

Figure 2.3—Capacity Data Flow

should also be known and managed at the training infrastructure level.

All these data flow to strategic training management, where resources are allocated and plans developed to bring the capacity-constrained training requirement (TPR_{Final}) as close as possible to the unconstrained requirement ($TPR_{Initial}$). These data are also used to help prioritize long-term capital expenses.

Finally, the corporate level needs the capacity data to project the production training and hence future manning levels. These projections help in the planning process to determine alternatives for complying with the Defense Planning Guidance (DPG) and for the analysis of trades among warfighting and funding options. The corporate level is also where the demand for training normally starts. The initial unconstrained training requirement is the number of trainees, by skill level, needed to replenish skilled manpower based on retention, authorizations, cross flows, other service training, and promotions. The final training requirement is developed based on the capacity data. Thus, a feedback loop occurs as capacity projections flow into the corporate level and requirements flow out to strategic training management. Refinements on this capacity data may also be required as complicated tradeoffs among student skill areas (the AFSCs) at capacity constraints are considered.

Quantity Information Data Needs and Flows

Training production data are the primary source of quantity information. Figure 2.4 shows how quantity data should flow in the system.

Production data are best known at the direct training level and are first tracked there. Quantity information is gathered to track students as they move in and out of training status, to assign class seats, and to project flow.

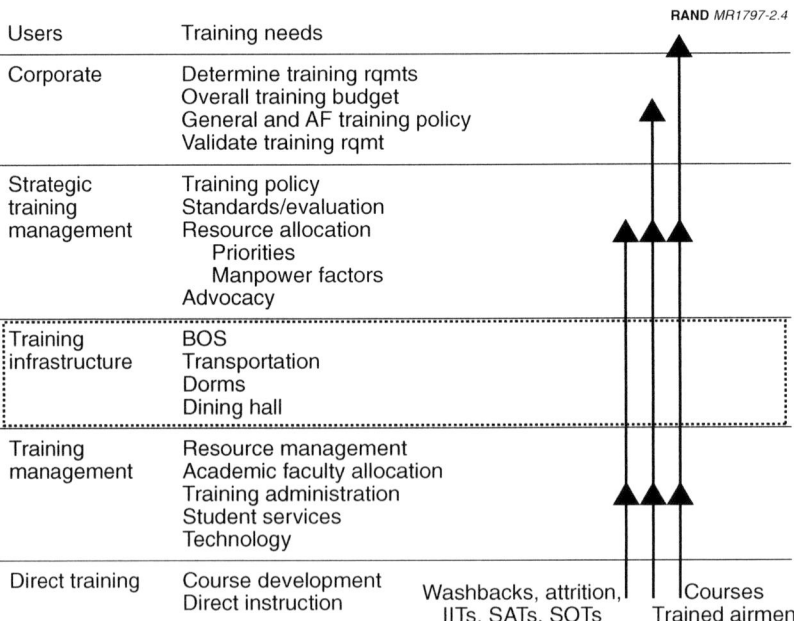

Figure 2.4—Quantity Data Flow

Training management should have access to all quantity data for the units it manages. It uses the data for resource utilization (e.g., in assigning dorm space to squadrons).

Strategic training management should access the entire production data set to make sure production by skill area is on target and to determine whether resources need to be reallocated. (It should also use the current production data to determine whether long-term changes are needed for additional capacity or recapitalization of current capacity assets, or whether new systems and missions are coming on line.)

Finally, the corporate level requires production data to evaluate users' requests for changes in requirements. The corporate level also uses the data to set goals for cross training and to assign production quotas for other users, such as the Guard, reserve, other services, and foreign governments.

Quality Information Data Needs and Flows

The direct training level uses quality data to evaluate how well students are being trained and the effectiveness of instructors, training devices, teaching methods, and supervisors. In some cases, the effects of resource shortages can also be inferred.

The training management level uses quality data to evaluate the effectiveness of the direct training units' leadership and to ensure standards are uniform among the schools. Training management can also use the data to evaluate environmental impacts on training—i.e., the effects of classroom type, class size, dorm quality, training day length, squadron leadership, and other, special conditions existing at the bases.

Strategic training management develops the overall standards and evaluation (Stan/Eval) requirements for all training. At this level, policy and rules are applied uniformly to all units. Quality data are used to evaluate the policy, rules, and standards across the entire technical training spectrum, particularly during the Utilization and Training Workshop (U&TW). This workshop is held approximately every three years for each AFSC to review requirements, curriculum, and training quality. Indicators of training deficiency or overall training quality may show up at this level.

The corporate level evaluates the effectiveness of the current training curriculum against operational needs. It projects and requests changes to existing training programs to meet current deficiencies and future requirements. Four major sources of quality data are accessible to the corporate level: (1) the Graduate Assessment Survey (GAS), (2) the Field Evaluation Questionnaire (FEQ), (3) field interviews (FIs) of graduates and supervisors, and (4) the Occupational Measurement Squadron (OMS) survey results.[11]

The GAS asks supervisors to judge the effectiveness of the technical training program. It is a short survey with only three basic questions.[12] The FEQ is also directed to supervisors. Its goal is to improve training, and it is very specific, containing many questions unique to the individual course being evaluated.

FIs are for both supervisors and graduates and have goals similar to those of the FEQ: to determine whether training is appropriate and in the right amount for the requisite tasks. FIs are a costly means of acquiring data since they require personnel to travel to bases around the world and interview assigned personnel. As a result, sample sizes are small.

The training extract of the OMS survey is used to assist in training decisions. The primary users of the data are the functional managers and attendees at the U&TWs. The training abstract measures, by specific task, the percentage of personnel (distinguishing skill levels 3/5/7) performing various tasks and the difficulty of the tasks. Numerous other breakouts of the data assist decisionmakers in revising the Career Field Education and Training Plan (CFETP).

---

[11]The OMS survey results do not measure graduate quality, but they are an important measure of training quality. Here, *quality* refers to training matching the needs of the Air Force.

[12]The three questions are as follows: (1) What is your assessment of the graduate's attitude and adherence to military standards? (2) How would you rate the graduate's ability to perform at the apprentice level as defined in the CFETP/STS? (3) How well do the apprentice job requirements outlined in the CFETP/STS meet the job requirements in your workplace?

Figure 2.5 shows the desired quality data flow. The process starts at the direct training level but depends on data that come from the users, individuals, and supervisors directly affected by the courses and training that AETC gives. The data flow to training management so that a review of each schoolhouse can be carried out, the concern at this level being the quality and relevance of the material, as well as feedback on instructor performance. Strategic training management, which uses the data to manage training quality, needs the data at a higher level of aggregation. Then, when the data finally reach the corporate level, they must be disaggregated so that the CFMs can evaluate the quality results for their individual skill areas.

SUMMARY

This chapter has presented a model training management structure with as little reference to current Air Force management practices as

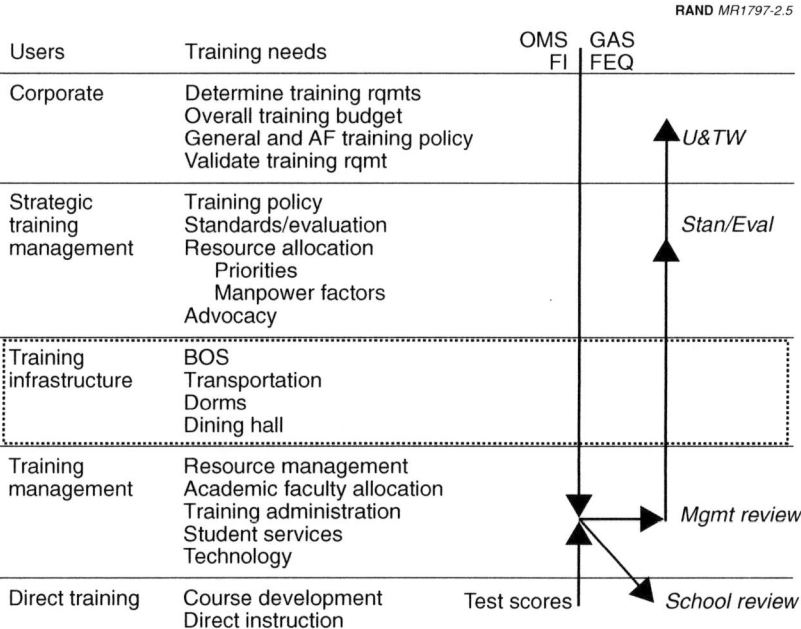

Figure 2.5—Quality Data Flow

possible. This model highlights decisionmaking responsibilities and the need for information to support decisionmaking. This in turn establishes the demand for data and the data flows.

We now turn to the current Air Force organizational structure and use this framework to highlight improvements that can be made.

Chapter Three

CURRENT AETC TRAINING MANAGEMENT
STRUCTURE AND DATA FLOWS

Having outlined an exemplary structure for training management and the associated flow of data among the functional levels, we now map the current Air Force organizational structure onto this model. We begin with the organizational mapping and conclude with a detailed examination of data flow deficiencies in the current system.

CURRENT AETC MANAGEMENT STRUCTURE

Figure 3.1 maps the different levels of technical training delivery and management to the specific AETC and other Air Force organizations. This mapping is based on our reviews of AETC documents and interviews with personnel at all levels of decisionmaking.

Starting from the bottom (where the majority of the data are created), we observe that for direct training, training management, and training infrastructure, there is not much of an overlap in functions among the current Air Force unit types. Each training squadron (TRS) is responsible for the direct teaching of a set of related classes; the training group (TRG) serves as the command structure for all TRSs at an individual base, and the training support squadron (TRSS) functions as the TRG staff in the training management level. In training infrastructure, most of the functions involve the TRG to some degree, except for the pure base functions, such as BOS, which is the responsibility of the training wing (TRW).

At the strategic training management level, however, decisionmaking (curriculum direction, capacity, future capabilities, etc.) is very de-

RAND MR1707-3.1

Users	Training needs	MAJCOMs		
Corporate	Determine training rqmts	DPR, U&TW		
	Overall training budget	DPR, XPM		DPL (new)
	General and AF training policy	DPD, XOO		
	Validate training rqmt	??		
Strategic training management	Training policy	AF/DPD, SAF/AQ, DO		
	Standards/evaluation	2AF,	TRG	
	Resource allocation	DO/XP		
	Priorities	DO/XP		
	Manpower factors	XP		
	Advocacy	AF/DPR, DO, XP, 2AF, TRW		
Training infrastructure	BOS	TRW		
	Transportation	TRW,	TRG	
	Dorms	TRW,	TRG	
	Dining hall	TRW,	TRG	
Training management	Resource mgmt	TRG, TRSS		
	Academic faculty allocation	TRG		
	Training administration	TRSS		
	Student services	TRSS		
	Technology	TRSS		
Direct training	Course development	TRS/TRSS		
	Direct instruction	TRS		

NOTE: DO and XP refer to AETC/DO and AETC/XP.   As of 30 Sep 02

Figure 3.1—Distribution of Technical Training Functions Among Organizations

centralized, with multiple levels of control that often overlap. Sharing these responsibilities are the Air Staff (AF/DP, AF/XO, SAF/AQ, and the CFMs), AETC/DO and AETC/XP, 2AF, and the wing level at each base. As a result, there is no single voice for training at the strategic training management level in AETC or at the Air Staff. Each organization has some responsibility for strategic decisions and demands data from the other levels to support its own decision processes. Also, having multiple communication and reporting systems with AETC leads to confusion in planning and execution cycles.

The result is that strategic decisions are apportioned to one of the three levels on an almost ad hoc basis. In some cases, decisionmaking responsibility is a matter of historical happenstance; in others, it is the result of a staff member having taken the initiative to assemble and report on data (which sometimes has left a gap when the staff member retired or moved elsewhere). This is directly reflected in the lack of reliable data-reporting channels and hence the need for ad hoc queries and sometimes staff resentment when they are faced with special data requests, especially for decisions they view as being in someone else's purview.[1]

At the top of the structure, the corporate level, there currently is no single office to trade off quality, quantity, and compensation initiatives from a training point of view.[2] Course content is determined in the U&TW, the members of which include the CFM, the MAJCOM users, and the training experts associated with the requirements for each AFSC. The Director of Personnel Resources (DPR) oversees the U&TW, which evaluates the effectiveness of current training against the requirements for each AFSC. DPR projects and requests changes to existing training programs to meet current deficiencies and future requirements; the U&TW works to ensure that training quality meets the users' requirements. A senior AETC officer chairs an issue resolution meeting, if needed.

Corporate also influences the specialties' training priorities. The initial TPR is created without regard to capacity or cost effects.[3] The corporate level is involved in at least three ways in order to balance training requirements with the training budget:

- *Budget prioritization.* Decisions are made as to which training programs will be emphasized (receive a greater budget share) and which will not. In the TPR conference (formally known as the Training Flow Management Conference), representatives from corporate attempt to integrate MAJCOM requirements

---

[1]In one example, notably 2AF, a special data system was constructed to connect TRS leaders directly with the 2AF/CC in order to ensure that some specific information was going to HQ 2AF.

[2]As this report was being written, the Air Staff was establishing DPL (Deputy Chief of Staff for Personnel, Learning), which combines a number of training-related offices. It is too early to evaluate the effectiveness of these changes because DPL is still developing.

[3]The initial TPR is a mathematical projection of requirements based on retention rates, authorizations, manning requirements, and cross flow. It does not consider cost and capacity. The overall TPR process includes cost and capacity but not in the initial formulation. See discussion in Chapter One for more details on the initial and final TPRs.

while simultaneously maintaining authorized force structure levels.

- *Offset funding.* When it is not possible to reallocate money within AETC's budget to fund a critical training need, corporate can adjust other, nontraining funding lines and increase the AETC budget for the critical training.

- *Requirements validation.* Corporate validates requirements, thereby determining whether specific requirements will ever receive funding.

The principal tool for these trades is the annual Program Objective Memorandum (POM) that includes AETC funding and other special training-related acquisition items, such as high-cost training devices. Because of this decentralization, data are developed at various levels and supplied in response to diverse requests to support budget drills and other asynchronous decisionmaking. This means that different data can be provided at different times because of the timing of specific requests, which leads to administrative confusion when the bases for decisions are compared. For example, 2AF, commanded by a two-star general, has no problem collecting capacity data from the direct training level. Yet AETC/DOO, the technical training division, led by a full colonel, has difficulty obtaining capacity data. We believe this fragmentation and overlap constitute a major source of the data problems that have been identified in AETC, although as we show below, other issues, such as methodological practices, are also significant contributors.

In summary, we believe that the flatter organizational structure discussed in Chapter Two, with a centralized strategic management function for technical training, is warranted.[4] While instituting an AETC/ADO (assistant deputy of operations)[5] with technical training expertise may help to centralize some of the data flow, our initial

---

[4]Essentially, we are defining the four broad levels as follows: HQ AF is the corporate, HQ AETC is strategic management, the TRGs are training management, and the TRSs are direct training.

[5]When we were writing this report, the creation of a separate ADO for technical training was proposed as a potential solution to some of these strategic management issues.

opinion is that it will not resolve all the organizational issues seen at the strategic management level.

DATA FLOW DEFICIENCY ANALYSIS

Having considered organizational levels and the corresponding data needs for decisionmaking, we are now ready to assess the capabilities of AETC's data systems.[6] We acknowledge that this assessment is inexact for a number of reasons. First, ad hoc data processing is still prevalent, which means that it is difficult to generalize across all data and all suborganizations involved in producing some pieces of the data. Second, we did not perform a comprehensive analysis across all of AETC's suborganizations; instead we interviewed HQ and 2AF leadership and staff along with wing, group, and squadron personnel at Keesler Air Force Base. Third, AETC's data systems are in a state of enhancement, so some of what we discuss here will likely be improved before this report is published. Therefore, what we present is a first-order assessment covering the flow of raw data, the processing of these data into information to better inform decisionmaking, and the organizational structure in which the data and information reside.

In evaluating the data flow, we made a distinction between data *availability* and data *accessibility:*

- *Availability* refers to the existence of (a) raw sources of the data in any form at some level within AETC and (b) the processing mechanisms needed to create information for decisionmaking.

- *Accessibility* describes the decisionmakers' ability to obtain the information necessary to aid in the decision process and is particularly sensitive to where and in what form the data reside.

If data are not available, they are by definition not accessible. Sometimes information is available but inaccessible because it is not disseminated in a manner that allows easy consumption. In other cases, inaccessibility is caused by data sources not being sufficiently

---

[6]We generally differentiate between (raw) data and information. Data usually require aggregation and manipulation with other data items to become information useful for decisionmaking.

recognized. More often, however, information is inaccessible because the processes or methods needed to combine raw data into information that could aid decisionmaking are deficient. For example, information on the total cost per student for a particular AFSC is unavailable because the methodology to appropriately allocate costs (such as for facilities and BOS) across AFSCs is deficient. (A more extensive discussion of a methodology for computing capital costs is in Appendix B.)

The following subsections describe each category in detail and explain what data are missing and why. We use a simple system of black, medium gray, and light gray to differentiate the deficiency: black represents no capability to meet the data needs, medium gray represents a partial capability, and light gray represents full capability. So, black under availability indicates that no data source for the information exists or that the necessary calculation methodologies for the information are unknown. Medium gray means that there is a known source for only part of the information and that either some significant portion of the data or some needed methodology for determining the information is unknown. Light gray means that the source and calculation methodologies are entirely known. Similarly, a black under accessibility means that mechanisms for disseminating the information are seriously deficient, medium gray indicates that the information can be obtained by special request or that a special effort is required to transform it to a usable format, and light gray means that the information is readily available in the desired format.

Cost Data Flow Deficiencies

Figure 3.2 shows our ratings of the deficiencies in the major categories of cost data at each organizational level.

Operations and Maintenance. Table 3.1 shows the categories of O&M costs for direct training. These costs are primarily expended and monitored at the group and squadron level (the training management and direct training levels).

To reduce costs and increase flexibility, recurring O&M contracts are typically pooled across TRSs (and sometimes TRGs) with no delin-

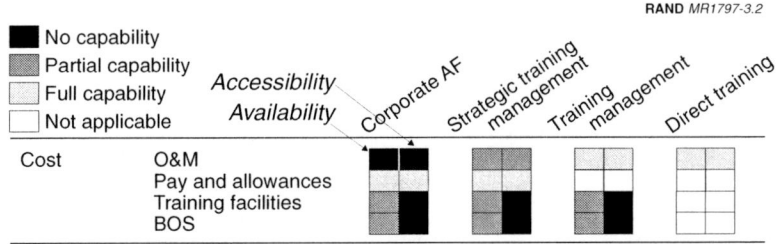

RAND *MR1797-3.2*

Figure 3.2—Cost Information Data Deficiency Ratings

Table 3.1

O&M Costs for Direct Training (Inside the Classroom), Including Direct Training Support

Recurring	Nonrecurring

• Equipment/devices/products— maintenance and repairs
• Consumables
  – Supplies
  – Petroleum, oil, lubricants
  – Replenishment spares
	• Equipment/devices/products—new development, procurement, and transfer
	• Initial spares/stocks
	• Data management system design, development, procurement, and upgrade

eation of specific work performed. However, this and similar data collection at this level can lead to aggregations across training units that are difficult or expensive to disaggregate (and aggregate).[7] The aggregate data are generally accessible to higher-level management but cannot be exactly divided up on a course-by-course basis. Costs for consumable supplies are generally associated with specific courses or squadrons and have been mostly accessible through manual systems, although TTMS can track them in an automated database. Air Force organizations are required to update the base accounting system with information on purchases. Most organiza-

_____

[7]These data have the potential to be useful if shared with lateral training management units.

tions run ad hoc programs and spreadsheets to track and project these costs. Typically, the ad hoc databases are more up-to-date than the base system is and are easier to use.[8]

On the nonrecurring side, groups or squadrons usually request capital or one-time investments in equipment, devices, products, spares, or other stocks, but upper management must approve such requests (including those for inclusion in the POM). These costs are usually tracked in a combination of automated and semi-automated processes. Capital items are sometimes procured directly from a MAJCOM or other source,[9] in which case AETC usually pays the recurring O&M costs but not the nonrecurring costs. Finally, the cost of data systems to support direct training and training management (including internal management systems and reporting/dissemination systems) must be included. In the past, most of these systems were ad hoc and peculiar to each organization. With AETC's TTMS and ADSS initiatives, more of these costs are accessible to the strategic management level.

Two primary deficiencies exist in this data area, the foremost one being the lack of specificity for the equipment actually serviced in the recurring O&M costs. It was reported to us that previous attempts to determine these data either identified costs that were excessive (because of high administration costs or because economies of scale had not been implemented) or limited the flexibility of the O&M contracts. As a result, it is difficult to determine how to split up these costs equitably among students to determine either the marginal or the average cost to train a student in an AFSC.[10] This problem particularly affects the corporate level. Because corporate managers also complain that the information is not readily available to them, we rate as black both the availability and accessibility of O&M data at the corporate level.

---

[8]Though the data exist in these ad hoc databases, they are not standardized and would not necessarily provide the information needed at the next level.

[9]For example, equipment may be obtained as part of the system acquisition process or from closed bases.

[10]This is further complicated by the fact that there are no defined business rules for sharing the O&M cost of training devices used by more than one AFSC.

The second primary deficiency is for the nonrecurring O&M costs. In this case, the methodology for combining the raw cost numbers with supporting data—such as student throughput, acquisition factors, equipment life cycle, and upgrade alternatives—is incomplete. These data typically flow to the strategic training management level in a semi-automated form, so we rate as medium gray both the availability and accessibility of the O&M cost-related data at this level. It appears to us that O&M-related data are adequately available and accessible for decisionmaking at both the training management and the direct training level.

Pay and Allowances. The pay and allowances data necessary for making decisions at both the corporate and the strategic training management level seem to be readily available and accessible, but their availability and accessibility depend on the status of MILMOD.[11] In this category of cost data, as shown in Table 3.2, diverse data need to be collected on civilian and military trainers, staff, and administration. Generally, these data are unnecessary for decisionmaking at the two lowest levels of management.

Table 3.2

Pay and Allowance Costs for Direct Training, Course Development, Evaluations, Course Administration, and Strategic Training Management

Recurring	Nonrecurring

• Civilian (direct training)
   – Pay/allowance
   – Replacement acquisition and training
 • Military (direct training)
   – Student PCS/TDY
   – Instructor/manager TDY
   – Replacement training
 • Group/support squadron staff
   – Contracts
   – Registrar
   – Administration
 • AETC/2AF/training wing administration
	• Civilian
	   – Acquisition for new course
	   – Training for new course
	   – Separation/transfer
	 • Military
	   – Training for new course
	   – Instructor/manager PCS

---

[11]MILMOD is the modernization of the Air Force personnel data system—its hardware, software, and architecture. MILMOD (or MILPDS [for Military Personnel Data System], as the database is called) is now on-line but is experiencing numerous growing pains.

It is significant that some of these costs are in specific funding categories (e.g., for TDY) that cannot be mixed with other funding categories. This can require some complicated financial management. However, these systems increasingly can provide information in an automated form. Often the cost data need to be combined with such data as manning, instructor turnover rates, and graduation rates. These data also seem to be readily available and accessible. For these reasons, we rate as light gray both availability and accessibility for the top two management levels.

Training Facilities and BOS. Finally, we turn to costs for training facilities—i.e., those buildings used exclusively for training, such as classrooms, dorms, dining halls, and student centers—and BOS costs, as shown in Table 3.3. A distinction is made between training facilities and BOS because all training facility costs are associated with training, but only some fraction of BOS costs should be, since most bases serve functions in addition to training. In either case, recurring costs include those for utilities (electricity, gas, water) and for maintenance of buildings and grounds. Transportation costs (some POL [petroleum, oil, lubricants], motor pool, and roadway maintenance) are usually included in the BOS costs. Under nonrecurring costs, we include new building construction and major remodeling, as well as the BOS cost for vehicle purchases and roadway construction.

Decisions based on these data are hampered for two major reasons. First, a comprehensive methodology for allocating these costs has

Table 3.3

Training Facilities and Base Operating Support Costs

Recurring	Nonrecurring

• Utilities and maintenance
  – Training facilities (classrooms, dorms, dining halls, student centers)
  – BOS (infrastructure such as BX, medical center, gym, post office, support services)
• Transportation (BOS)
  – POL
  – Vehicle and roadway maintenance
	• Construction and remodel
	  – Training facilities
	  – BOS
	• Transportation (BOS)
	  – Vehicle procurement
	  – Roadway construction

not been developed. AETC seems to have access to most if not all of the raw budgeted costs for these capital assets, since they are used to justify POM budgets and prioritize MILCON projects; but these raw costs must be converted to useful support information for decision-makers. Currently, these costs are included as much as possible, sometimes with such additional information as student throughput, facility utilization factors, facility life cycle, and student projections by AFSC. Part of the difficulty in establishing a methodology is how to deal with uncertainty about the future. And some of the utilization factors may need updating. For these reasons, we rate availability as medium gray for all appropriate management levels.[12]

The accessibility of this information is the second major problem. Construction data are handled under MILCON budgets and, when accessible at all, are only available manually. We see a general need to improve data accessibility in these areas, so we rate accessibility across all levels as black.[13]

In addition, there is a question as to why AETC needs accurate cost numbers. We identified three reasons: to justify POM budgets, to charge other training consumers, and to make efficient production decisions. For these purposes, AETC needs both average and marginal costs. Average cost divides total costs—which include both variable costs (for training materials, instructor salaries, etc.) and fixed costs (for dorms, classroom buildings, etc.)—by number of students. While average cost is appropriate for justifying budgets or charging foreign governments, production decisions rely upon marginal costs. Marginal cost—the additional cost of producing one more unit—treats fixed costs as irrelevant to current decisions, since the costs are sunk as a cost of doing business. These additional methodological points need to be considered to ensure that accurate supporting cost data are provided to financial decisionmakers at AETC. (See Appendix B for a more detailed discussion.)

---

[12]It appears to us that decisions in direct training require neither training facility nor BOS costs.

[13]Wings tend to have more insight into training facility and BOS costs because of their organizational responsibilities. Even so, this is an area where improvements are needed.

Capacity Data Flow Deficiencies

Figure 3.3 shows the deficiency ratings for each type of capacity data information at each organizational level. The following subsections discuss the ratings in detail for each type of capacity information.

Classrooms. Classroom space and utilization are key in determining optimum class size and are required to plan for class flow.[14] They are of most interest to direct training and training management, because these two levels schedule classes and decide whether facilities are too crowded, are underutilized, or require upgrading because of age or technical obsolescence.

As the primary provider of resources for major changes in classroom space, strategic training management presumably could use this information to help inform its decisions. But because there are multiple classes and instructional facilities, the detailed data from the direct training level most likely need to be organized into some form, such as a model, that will allow strategic training management to explore the consequences of decisions. Data on overall classroom utilization should be sufficient for strategic training management, and the level of aggregation needed for this level also seems appropriate for classroom information at the corporate level.

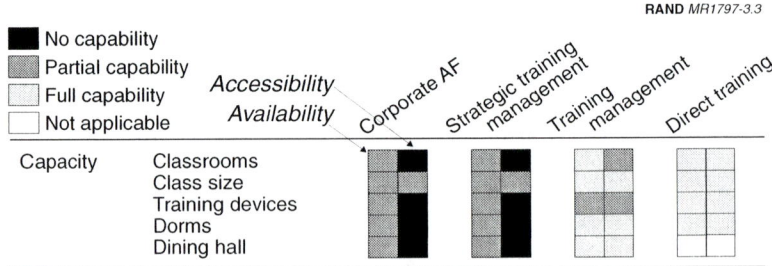

Figure 3.3—Capacity Information Data Deficiency Ratings

---

[14]The Training Planning System (TPS) maintains information on minimum, maximum, and programmed group size. TPS does not provide information on classrooms or utilization of classrooms.

Direct training has control of its classrooms and knows their capacity (seating, equipment, etc.). Classroom data do not appear to be kept electronically, although some units may do so under local initiatives. Training management does not maintain classroom information internally but can request it from the squadrons if needed. Currently, strategic training management does not maintain classroom information in any form and would have to request it from training management.

Class Size. Current class size is one of the key operating parameters for the AETC system because it feeds into decisions about recruiting needs, future available airmen with specific AFSCs, and more-local measures, such as students awaiting training (SATs). Further, managers need to know the maximum (surge) capacity and the optimum capacity[15] of each class and class segment. Note that this is a (possibly complicated) function of instruction facilities and instructors. These numbers need to be available to all levels of management, although the need at the corporate level is for aggregate, yearly numbers, not the capacity or size of individual classes.

Of the four management levels, direct training obviously knows current class sizes and has the most direct information on both the maximum and the optimum class size. Direct training maintains paper and electronic records (primarily paper concerning optimum class size, although some information may be personal knowledge carried by the instructors but easily available). At the training management level, class size should be available via the training management registrar; class capacity information is available from the TRSs directly. Strategic training management gets information sporadically and has to request specific information at some levels. Corporate apparently receives this information only sporadically and at long intervals. We observed that corporate needed to request it from AETC.

Training Devices. Many of the more technical AFSCs use a variety of training and test devices to give students hands-on experience with

---

[15]Both the maximum and the optimum are needed to ensure that the maximum is not made the normal class size in pursuit solely of maximum throughput. In practice, the optimum is always set at the maximum, a practice that discourages the development of realistic surge factors.

relevant parts of equipment they will be working with in the field. In many cases, these devices are facsimile components built locally so as to avoid the need for actual copies of operational equipment. For replacement and planning purposes, training management needs to know rough cost and what class(es) the device is used for, and the device must be documented (if home built) because the training system must be able to repair/rebuild/replace the device to maintain training efficiency.

In all cases, direct training knows first-hand most of the details of the individual training devices used by the course, but it does not currently keep a "centralized" list, nor is there a formal process for ensuring that the equipment is documented. Many units use ad hoc databases since there is no central database or accepted format for saving such data. Training management does not keep records either and would have to poll each squadron individually.

Strategic training management does receive information on training devices in course resource estimates (CREs), which are generated by the TRG and go directly to AETC/DORB as course changes are requested. Otherwise, strategic training management has no visibility on the number of training devices, their lifetimes, and information about replacement needs and costs. It can request these data from direct training via training management.

Neither does corporate receive information on the number of training devices. Currently, corporate does not request this information, but it is not hard to imagine capacity-related trades for which this information would be useful.

Current plans are for TTMS to maintain information on training devices above a certain cost as part of the curriculum/syllabus information.

Dormitories. Dormitories straddle the line between direct training and training management. Each dorm usually houses students from different classes, but the military side of direct training tries to place students in such a way as to maintain group integrity. Gender separation complicates this goal and leads to compromises. Given these imperatives, direct training and training management need to know all the details of dorm space and configuration. Strategic training management requires information on total dorm capacity. It also needs information from training management about specific problems with configuration so that it can develop surge plans.

Direct training and training management know dorm capacities and configuration details because they assign and manage dorm space on a week-to-week basis. However, the exact form of this information is idiosyncratic to individual bases. Currently, dorm information goes to strategic training management on an ad hoc basis and does not seem to be monitored regularly.[16] Some information is provided to strategic training management via AETC Civil Engineering databases that draw from each base, but this seems to be total building information (e.g., square feet, building condition). It is not clear that the corporate level needs any of this information (e.g., numbers of individual rooms), except via strategic training management in planning for housing capacity adjustments.

Dining Facilities. Dining facilities feed the entire student population, and their primary characteristics are the total capacity and the number of students that can be handled per "wave" within mealtime windows (mealtimes are staggered so that different classes arrive at different times to smooth the flow). Direct training and training management need to know the capacity for operational purposes, primarily to schedule classes and specify base transportation needs.

Some bases contract out the food service, in which case the base contracting office (as part of training infrastructure) maintains the capacity information. It can be argued that because the wing is involved, strategic training management has some information, but the resource-allocation parts of the fragmented strategic training management structure (especially at HQ AETC, where budgeting is done) do not have direct information. Like dorms, dining facilities may be a significant constraining factor. As is the case with dorm space, corporate seems to have little need for detailed information on dining facilities.

---

[16]In one case, the 2AF/CC took a personal interest in why A-10 maintenance production was falling short, resulting in the identification of dorm constraints and the subsequent building of a new dorm.

Quantity Data Flow Deficiencies

In the area of quantity data, much has changed in AETC over the past three to five years. AETC has made a significant effort to develop systems to better capture and manage production. Two primary systems are TTMS and the AETC Decision Support System (ADSS). TTMS, the primary data warehousing system for AETC, is coming online with some parts operational, and each quarter brings progress toward a more comprehensive system. ADSS is an interface system for the data so that they may be processed and aggregated as desired. Additionally, the Air Force's MILPDS is useful for linking demographic data and creating longitudinal information with TTMS.

Unfortunately, errors in the production data persist. Some errors stem from continuing problems with MILPDS, which is undergoing large-scale changes. Others stem from the training data input to TTMS. Additionally, units in the past tracked production using ad hoc systems and the Air Force Training Management System (AFTMS), which then links to MILPDS. However, MILPDS does not contain the standardized reports that were part of the foundation of AFTMS, so creating the desired reports and checking the data's veracity are problematic. In time, as the causes of errors are better understood, procedures and policies should be put in place to eliminate most of these problems.

Figure 3.4 shows the quantity data deficiency ratings.

Production. Due to the steady maturation of TTMS, the availability and accessibility of production data are good. Errors in these data and interoperability with MILMOD are the only areas of concern.

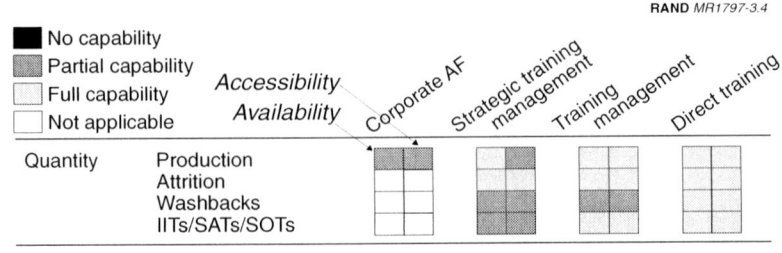

Figure 3.4—Quantity Information Data Deficiency Ratings

These data are used daily at the direct training and training management levels. ADSS is used to access the data at the strategic training management level. Some offices at that level and at the corporate level observed deficiencies when trying to obtain specific cuts of data at the right aggregation levels. But overall, production is the data category with the best current flow in the training system. Corporate has access to production data, but our conversations with action officers at the corporate level revealed that they were unaware of this access.

Attrition. At the direct training and training management levels, attrition data are tracked through TTMS. ADSS aggregates attrition data for metric use (evaluation of actual results meeting planned results) at the strategic training management level. Corporate uses attrition data in developing TPR goals. Attrition is a major component in calculating the cost of training and in determining the required schoolhouse inputs of students. Thanks to the implementation of TTMS and ADSS, the availability and accessibility of attrition data are good across all levels of management.

Washbacks. Washback data are available at the direct training level, but some issues arise at the training management and strategic training management levels concerning a standard definition for washbacks. Some airmen wash back more than once, so the question is, Should washbacks be defined based on the number of people who wash back or the number of times individuals have washed back? A common business rule defining how washbacks are measured would solve this issue. The data are currently recorded in TTMS.

IITs/SATs/SOTs. Student status is input and tracked at the direct training and training management levels through TTMS. At the strategic training management level, some of the quantity data on students ineffective in training (IITs)[17], students awaiting training (SATs)[18], and students out of training (SOTs)[19] are difficult to obtain

---

[17]Students who, for medical, behavioral, or personal reasons, are not currently in a training course.

[18]Students who could be in a course but are waiting for their next course to start.

[19]Students who have completed their course work and are waiting for their first assignment.

because they are tied to individual records and raise issues of privacy. This issue could be fixed by clearly defining the business rule to correctly aggregate the data by courses and AFSCs.

Quality Data Flow Deficiencies

As noted earlier, test scores and a variety of surveys sent to students' supervisors measure quality. Test scores are recorded for each individual student in TTMS. Students must earn minimum test scores for graduation and in some cases certification.

Figure 3.5 summarizes our detailed assessment of the availability and accessibility of quality data to meet the needs of decisionmakers. The majority of quality data does flow smoothly through the training system and are available and accessible to the appropriate levels. For the most part, the direct training and training management levels have the data to answer questions and make decisions.

Test Scores. Test score data are available to the direct training and training management levels through TTMS. Test score data are not needed above the training management level.

GAS. The Graduate Assessment Survey (GAS) is Web-accessible to all levels of training. It provides multiple ways to view and aggregate the data. The information is useful and sufficient for all management levels. The CFM, at the corporate level, has access via the Web.

FEQ. The Field Evaluation Questionnaire (FEQ) data are in the process of becoming available through the Web. FEQ data are stored at

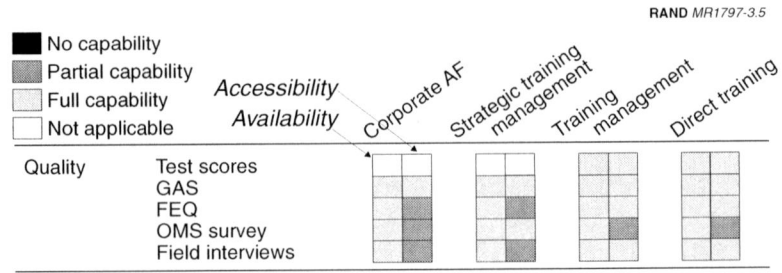

Figure 3.5—Quality Information Data Deficiency Ratings

the training management level and regularly briefed to the direct training level. The CFM, at the corporate level, has access to paper copies.

OMS Survey. The Occupational Measurement Squadron (OMS) survey data are developed at the strategic training management level and flow to all levels via a paper document. The data are clearly available; but because the documents are large (often 200 pages or more), the data are not very accessible to the direct training and training management levels. The data are primarily used in the U&TWs by the corporate level.

Field Interviews. Field interview data are held at the direct training and training management levels. These data are available to strategic training management only upon request and only in hard-copy format, so we rate them lower for accessibility.

In general, it is not clear that the quality data are being used by any organization (within strategic training management and corporate) other than the CFM during the U&TWs. Currently, the corporate level primarily uses the OMS survey data in a paper format, updated on a three-year cycle. The U&TW tries to ensure that training quality meets the users' requirements. The most likely organization to use quality data (at the strategic training management level) to identify potential training deficiencies is AETC/DOO, but it has too few people to follow up on the large number of enlisted AFSCs and courses.

Direct training and training management regularly use quality data to evaluate the effectiveness of training. For example, in the 37th TRG, a staff is assigned to regularly track and report quality trends.

DEFICIENCY SUMMARY

Figure 3.6 summarizes our findings on the availability and accessibility of the four classes of information. The end goal is to provide useful information to decisionmakers on the pipeline's cost, capacity, quantity, and quality. In some instances, solving the data collection or data flow problem would not resolve the data deficiency at a given management level, since raw data must be manipulated into a useful form for decisionmaking. Several methodological problems currently block the creation of cost information that would be ac-

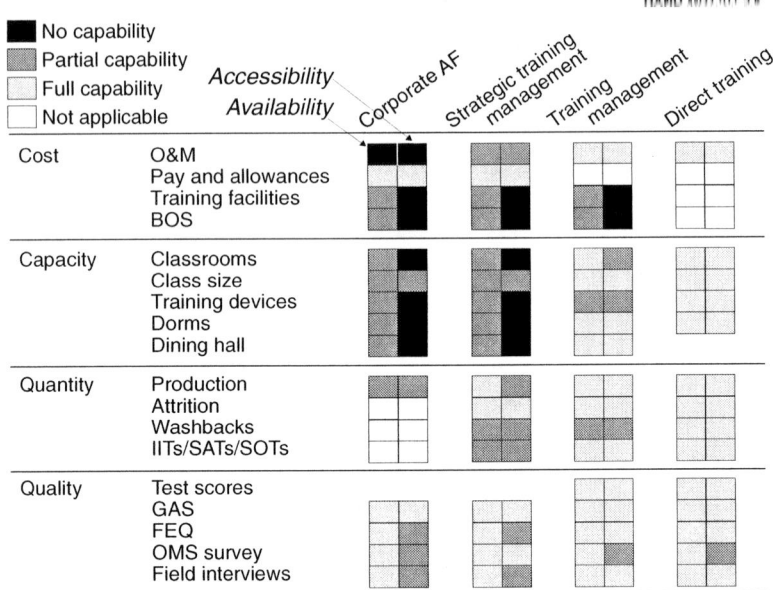

**Figure 3.6—Summary of Deficiencies by Data Type
and Organizational Level**

tionable by decisionmakers. Our overall assessment is that although better data automation will certainly resolve some of the information problems, the more difficult problem is designing the methodologies to transform raw data into information useful to decisionmakers.

Beyond the methodological issues, our interviews and review of AETC's organization and procedures revealed that many of the relevant data for doing cost and capacity estimates are available somewhere in the system, often at the direct training level. The level where most deficiencies are seen is strategic training management. We believe that some of the data flow problem results from there being no single focal point for technical training at the strategic training management level. TTMS and ADSS have particularly improved the collection of quantity data. We believe that a central data warehouse for cost and capacity data is also warranted, but that a new system for collecting the data is not needed—the data are already being collected. A system is needed to increase the accessibility of the right kind of data to the strategic management level.

NEED FOR REAL-TIME DATA

Does AETC need a new real-time data management system? AETC staff and decisionmakers have both asked this question.

First, the term *real-time data* requires a definition. In an industrial production system, *real-time* means a rapidly updating system that provides both an accurate snapshot of current operations and sequences of these snapshots (with time gaps between sequences that are as small as or smaller than significant changes in the system status). In a manufacturing process, the time between snapshots can be as little as fractional seconds because of the rapid production time of the subassemblies. What is *real-time*, then, for AETC? Certainly, the operational snapshot needs to be accurate based on the time it was taken. But does AETC require a data system that updates as rapidly as a manufacturing assembly system?

For example, three of the four management levels want real-time information on production and class capacity in order to track and detect washout/washback problems.[20] The gap duration depends on what the management levels want to do with the data. In the short term (on the order of weeks), it seems likely that washout/washback is a problem of individual students and should be dealt with at either the direct training or the training management level. In this case, gap time does not need to be nearly as short as for a manufacturing process. Upper-level organizations need to intervene only when these problems persist at fairly high levels over several different class cycles, which allows an even wider gap time. Even then, however, at least some of the causes (such as instructor problems) would still be first detected and dealt with at the lower levels (training management or direct training). Generally, status of training data may only change on a weekly or monthly basis. Therefore, so-called real-time data for AETC—and particularly for strategic

---

[20]ADSS, using data from TTMS, can provide real-time information on production only. The TPS provides class size data. No single system or database brings these various sources together.

training management—must provide accurate data snapshots but need not be updated at the minute-by-minute rate used for real-time manufacturing systems.

CONCLUSIONS

Clearly, deficiencies exist in the flow of cost and capacity data in the Air Force's technical training system. We believe that these deficiencies reflect a systemic problem consisting of both organizational and data flows. The lack of a single office and voice for technical training at the strategic training management level hampers the collection of data. Organizational issues may well represent over half of the problem of collecting cost and capacity data. Methodologies are lacking to properly account for certain types of aggregate information, particularly cost. Finally, a central data warehouse is needed to record and collect cost and capacity information.

Chapter Four

FINDINGS AND FUTURE WORK

In the main, the training provided by the U.S. Air Force (largely through AETC) is widely praised. The technical training that the enlisted corps receives, both initially and as continuing education, is considered excellent preparation for a wide variety of related civilian jobs. This excellence in direct training is the result of long-term investment in training, in training infrastructure, and especially in people at the direct training and training management levels. The relevance of that training to the needs of the field has been maintained by close contact between direct training and Air Force units via rotation of instructors, military trainers, and training commanders, and via several forms of feedback on student performance that go from the field directly back to the classroom.

However, there are pervasive concerns among training managers and personnel concerning the decisions and actions that take place above the level of day-to-day instruction management. Strategic training management must be able to deal with issues such as

- Widespread changes to the national security environment and the need to be rapidly responsive to them.

- The requirement for increased technological sophistication in many career fields.

- Technology opportunities that could enhance training.

- The need to balance and set priorities for the expansion or contraction of existing career fields.

- The need to swiftly address Infrastructure constraints (such as dormitory space).

- The need to perform all these operations within a tightly constrained set of resources, which requires that training be prioritized within the corporate Air Force and funds be allocated among AFSCs within AETC.

The Air Force is therefore less confident that it has the decision processes and the requisite data to support strategic decisionmaking.

This chapter summarizes our conclusions about the need to consider improvements to data flows and processing to support decisionmaking. But more importantly, it summarizes our findings about the need to define the roles and functions of training management. We then conclude with a discussion of future work needed to help improve training management.

FINDINGS

Importance of Training Management

AETC's perception of the strategic training management problem is that it is caused by a lack of data, particularly cost and capacity data, or at least by the inadequate flow of these data to the strategic training management level. This is one of the primary rationales for the current emphasis on completing AETC information system developments, mainly TTMS and ADSS, with their adjuncts, such as the AETC Master Status Panel and the AETC Command Metrics.[1] These developments are based on the implicit assumption that the real need is to collect and make available to the AETC commander and HQ AETC "real-time" training data, meaning current and very detailed data on a day-to-day basis.

We argue in this report that, especially with the fielding of TTMS, the problem is not the availability of data (with one important exception), but the accessibility of data in a form suitable for use by

---

[1] The AETC Master Status Panel and AETC Command Metrics Web pages are primarily tools to measure actual production against goals.

strategic training management.[2] We first summarize our findings on the availability of cost data. Then we turn to the general data accessibility deficiencies, which are strongly influenced by the fragmentation of the strategic training management function.

Availability of Cost Data

Most of the fundamental data on training capacity, quantity, and quality are currently being collected at the direct training and training management levels. The exception is cost data. In some cases cost data are not available; in others they are very difficult to obtain. This is partly because retaining and analyzing cost data is not a primary focus of the financial management staffs, which have been reduced through waves of downsizing: these staffs must concentrate on execution and putting together a budget for the next budget cycle. In our interview with HQ AETC staff, we learned that even when cost data are collected, there is not enough manpower to fully utilize them.[3]

A more serious problem—one that we learned from our interviews with the TRSs and all the way up to the senior levels at HQ AETC—is that methodological issues arise in trying to match cost data with individual AFSCs. Major facility construction is largely funded by the MILCON budget, which is not even directly under the control of HQ AETC, so facility costs are generally hard to get. And even if they were available, the lack of consensus on how to compute the time value of money in an appropriations-based budgeting system makes it difficult for AETC to amortize facility costs and plan timely replacements. Furthermore, allocating facilities shared among training units to specific AFSCs is intrinsically hard. Research is needed to help estimate and allocate the costs in this area. Appendix B discusses these issues in more detail.

---

[2]TTMS will improve and is improving the accessibility to quantity data.

[3]Reductions in the manpower available to perform cost analyses have occurred to respond to fiscal constraints and in the hope that computer systems would automate much of the process. Unfortunately, as noted, the automated systems have not always been able to fully respond to the need.

Accessibility of Cost, Capacity, Quantity, and Quality Data

TTMS should enhance and standardize the data on cost, capacity, quantity, and quality as they are fully fielded. TTMS should also make it possible to build databases that can hold many years' worth of such data.

However, for strategic training management, such micro-level data need to be analyzed at appropriate levels of aggregation. For example, in an educational system such as AETC, with course lengths ranging from weeks to months, *real-time* means analysis based on aggregating data from several course cycles. This implies that analysis techniques and modeling are required to show longer-term trends and explore the implications of alternative policy changes over a course of months and years.

Simple data warehousing (long-term storage of detailed data) may be useful for archival purposes to back up ad hoc databases from direct training or training management. However, this form of data is of little use for strategic training management. An analogy would be a collection of partially manufactured subassemblies and raw materials that require significant processing in order to become a final product usable by strategic training management. Because of the need to combine data across such categories as schools, time, and budget categories, the amount of disaggregated and descriptive data is likely to cause a scale problem. By this we mean both the amount of storage space required to hold these data and the amount of time needed to combine and manipulate the data to make them useful for decisionmaking. Business rules also need to be embedded in the data warehouse, along with the flexibility to extend the business rules and to construct ad hoc queries and combinations. Data warehousing generally seems to be more of an engineering effort than a science. However, for a single warehouse or a hierarchical set of warehouses to be fully supportive of AETC's management levels, we believe that significant effort must go into defining a system that is sufficiently extensive and extensible.

Organization of Strategic Training Management

As noted in Chapter Three, the strategic management functions for Air Force training are carried out in a number of organizations, including the Air Staff, HQ AETC, 2AF (one of AETC's two numbered Air Forces), and the individual wing organizations at each base. This fragmentation is largely due to historical reorganizations in the Air Force as a whole. The result is that data requests and decisions overlap, with some confusion as to who needs certain data, who should get them, and who should be responsible for manipulating and maintaining them. There is an even more fundamental confusion about who should be interested in specific issues and decisions. We have argued that no single organization is serving as the central strategic management for technical training, a role that would include surge planning, TPR harmonization, overall quality standardization, capacity analysis, and development of planning costs.

In contrast, the case studies described briefly in Chapter Two (and more thoroughly in Appendix A) for other services and industries show a different situation. In these organizations, there is a single point of contact for technical training. For example, in the Army and in the commercial organizations, training may be divided into functions (engineering, finance, artillery, etc.), business areas, or even geographic regions, but there is always a clearly defined senior person responsible for organizing the training and making the strategic decisions. This person has a staff for management and ultimately employs the training professionals who do the teaching. This also means that there is a focus for data development, with little possibility of conflicting lines of command or data demand.

This clear definition of responsibility for strategic management of technical training is missing in the Air Force. We believe that many of the problems with data availability and accessibility could be solved if this responsibility were clearly assigned to one person. Reducing the organizational levels, as we have suggested, will reduce the number of people the data must pass through. Furthermore, having the remaining levels more clearly focused on specific training management tasks will provide a better idea of who has source data, who has the responsibility to input and process those data, and who needs to receive the aggregated information for decisionmaking. Our example from AFMC demonstrates that this kind of organizational flattening is possible.

A number of equally good ideas could achieve this purpose. One option would create a two-star position for technical training (TT) on the HQ AETC staff and require reconsideration of 2AF's function. The TT would be dual-hatted with the 2AF commander, much as the Air University commander also fills the position of AETC/ED. All relevant staff from HQ AETC and 2AF involved in strategic management of technical training would then be under the direct command of AETC/TT, removing many of the organizational disconnects that we have noted. Training base wing commanders would report to the TT. This structure would closely align AETC's technical training strategic management with its function, thereby facilitating management functions and information flow.

As we were writing this report, AETC was considering two assistant deputy of operations (ADO) positions, of general officer rank—one to work technical training issues and one to work flying training issues. Creation of these two positions should help, but it is not clear that it will satisfy what we see as a need for a single technical training focal point at the strategic training management level. The key is not just organizational; the roles and process that each organizational level fulfills must be consolidated.

FUTURE WORK

As our study progressed, it became apparent that major research in the areas of training capacity constraints and cost calculation needed to be explored. To this end, we have begun designing simulation tools to examine capacity constraints and perform economic analysis to better deal with capital costs.[4] These design efforts are as follows:

1. Simulation of a technical training schoolhouse with explicit resource constraints that uses activity-based costing for planning purposes.

---

[4]Computer simulation of training is not new to AETC, but previous efforts were not sustained. RAND developed two training models in the 1970s (see Appendix A), and AETC documents mention a few internally developed models. All of these seem to have disappeared. Currently, AF/DP uses models to identify the training resources required for each AFSC. AF/DP also uses a Rated Management model, though the emphasis is on absorption, not training throughput. AETC/DO uses a fairly complex spreadsheet model to determine flying capacity, but it does not capture the uncertainties that a simulation can. There has been little recent work in simulation analysis tools for technical training.

2. Simulation of the end-to-end training process from recruiting through basic training, technical training, and on-the-job training (OJT) to determine pipeline constraints and to compare costs across the system.

3. Alternatives for applying capital costs to individual courses and students.

We highlight these simulation tools below, describing the work so far, along with what needs to be done in the future.

A Technical Training Schoolhouse Model

One planning tool that is missing at the strategic training management level is a comprehensive simulation of a technical training schoolhouse. A simulated schoolhouse can help identify constraints and the marginal costs of increasing capacity. RAND has started work on a model of a single schoolhouse. Figure 4.1 shows a schematic of an early version of the model.

The latest version of the model includes number of instructors, class size, assigned dorm capacity, dining hall capacity, number of classrooms, number of training devices, availability of training devices, shift policy, weekend policy, and syllabus sequence. It also includes the costs associated with facility use (dorms, dining hall, classrooms), instructor pay, consumables, and training device use. The goal is to develop the cost of producing an airman in a particular AFSC and to be able to say what contributes to that cost.

An End-to-End Training Model

Unfortunately, the cost of training and the impact of capacity constraints are not simple to compute. Policies and decisions made at

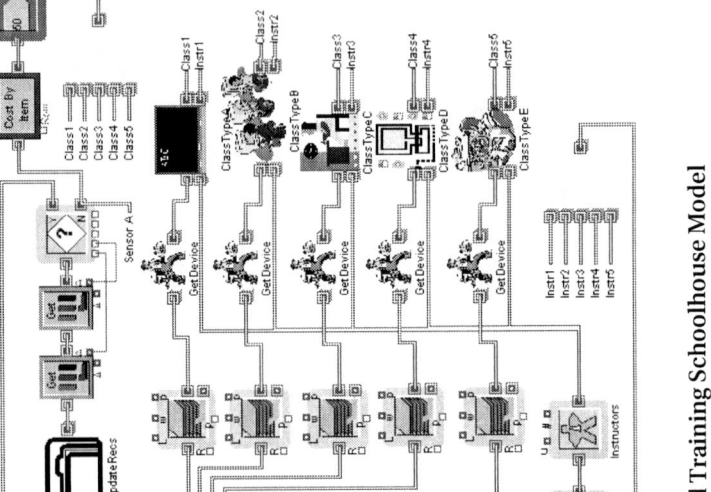

Figure 4.1—Technical Training Schoolhouse Model

the schoolhouses affect basic military training (BMT), recruiting, and OJT. Furthermore, changes in training policy (e.g., to improve the quality of graduates) may have a dramatic effect on both the format and the substance of courses within the schoolhouse.

Work has begun on an end-to-end training model to capture and articulate the effects of strategic training management and policy choices. Figure 4.2 is a pictorial representation of this model. Within each of the model's major areas are submodels that can track the flow of airmen through this system. Various measures may be captured—for example, the various cost measures shown in the figure. Such a model could also be used to explore interactions between schoolhouses.

The value of this tool can best be described using an example. Currently, one training base has set a policy to reduce students awaiting training (SATs) to zero.[5] To reduce SATs, the number of seats set aside for washbacks must be large enough to handle most outcomes. As seats are set aside, however, class openings for new recruits decrease. This causes qualified recruits to be left in the Delayed Entry

RAND MR1797-4.2

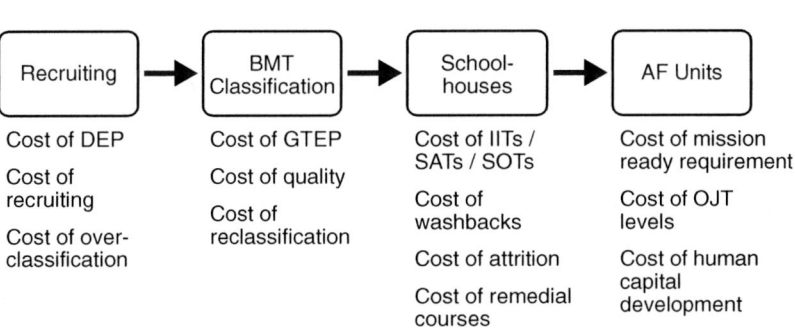

Figure 4.2—End-to-End Model Schematic

---

[5]We mention this example because one of the stated reasons given by the wing for its role is the need to set a vision for the training group. HQ AETC has specific goals for SATs (three days), SOTs (seven days), and IIT (six days).

Program (DEP), which means they must wait longer for class openings before being released to BMT. The longer a recruit is in the DEP, the higher the probability that he or she will not enter the Air Force. A recruit who leaves from the DEP represents some "lost" cost of recruiting. Is that cost greater or less than the cost of one SAT, or of a washback who has to wait for a class seat? And how does a zero-SATs policy affect maximum production? Clearly the model cannot answer the morale issues associated with a student having to wait a long time for a class, but it can show the cost effects that policies have on different portions of the training pipeline.

An end-to-end model could be very useful in numerous other analyses as well. It may be possible to combine it with RAND's work on productivity[6] to analyze tradeoffs between training at the schoolhouse and training on-the-job to determine the most cost-effective training strategies. We could also use the model to study surge options that depend on capacity constraints in recruiting, BMT, and technical training, and in the field (absorption).

Applying Capital Costs to Courses and Students

Our third effort concerns how to apply the capital costs of training to individual courses and students. The government is a unique business organization because money is not carried over between fiscal years. Budgets are developed yearly, and Air Force organizations are in a use-or-lose situation with unspent money near the end of the fiscal year. Commanders know that if they do not spend all budgeted money in one year, they risk receiving a smaller budget the next. This perverse incentive deters commanders from cutting costs and saving money for their organizations the next year. Government monies focus on the execution year, effectively yielding no time value to money within government. Organizations cannot borrow against future funding, nor can they save current money for future spending. AETC voiced this tough problem during an interim briefing in May 2002 and asked RAND to address it.

---

[6]The results of this work are reported in S. A. Oliver et al., *An Analysis of the Cost and Valuation of Aircraft Maintenance Personnel*, AFLMA, LM200107900, Jul 2002.

The capital cost dilemma breaks down into two distinct types of problems. The first is dealing with the here and now. When a capital resource is required but cannot be obtained (e.g., a space in a dorm that is completely occupied), what can be done? Often, instead of using this type of problem as motivation for pushing forward on a capital item, a short-term, more costly solution is found (e.g., using O&M money to buy hotel space off base). The second type of problem is finding a methodology to establish business rules for amortizing large capital costs into a yearly cost of capital. Some preliminary ideas for creating a methodology for distributing capital costs annually have been discussed, but much more work is needed in this area. (See Appendix B for a more thorough discussion of these issues.)

RAND will continue to look for ways to improve strategic training management, including the challenges it faces in terms of collection and the methodological problems associated with accurately attributing costs to individual students and capacity constraints.

LITERATURE REVIEW AND CASE STUDIES

Our literature review focused on work in five categories: previous RAND Corporation research, previous Air Force Human Resource Laboratory research, official (both historical and policy) Air Force documents, business literature on cost and capacity systems, and literature on organizational design and management structure. The results from that review are described in the following section.

We also conducted case studies of the management structures of training organizations within the other services and in the commercial world. These are described in some detail in the second section of this appendix.

LITERATURE REVIEW

Previous RAND Corporation Research

Mooz presented a paper at the Fourth DoD Cost Research Symposium in Washington DC in 1969, describing his study's purpose as follows:

> [T]o produce tools with which to analyze the pilot training process of the Air Force in terms of the resources required to train pilots and the cost of pilot training. . . . [T]he tools developed consist largely of mathematical simulation models which can be exercised on a computer.[1]

---

[1]W. E. Mooz, *Pilot Training Study*, RAND Corporation, Santa Monica, CA, P-4070, Apr 1969, p. 1.

His PILOT model simulated the entire pilot career track, from precommission through undergraduate and graduate pilot training and on into a potential 20-year career that included "desk jobs," cross-training, and nonflying status. The model specifically included variables for multiple types of resources and costs. The report contains examples and an analysis of a number of scenarios.

Boren built a model, called the Advanced Pilot Training Computer Cost Model (APT), in response to the Office of the Secretary of Defense's (OSD's) formation of a Pilot Advisory Committee to study "Pilots as a National Resource."[2] The report is one of eight volumes describing the use of the model. The last volume lists the FORTRAN code, about 600 lines.

Haley designed a technical training model that was to "be used for long-range planning, determining the effect of flow changes on training, [and] providing inputs for cost and resource analysis of training."[3] Haley developed a simple JOSS program as an aid in determining which policies and other factors should be in a technical training model.

In 1970, Hammond published a document covering part of an "initial phase of RAND's work in Air Force Technical Training."[4] This study focused on how to use technology to aid in the design of instruction for technical training. The report looked at a number of different techniques—such as mathematical techniques, simulation, and input/output modeling—and their usefulness in production planning.

In the late 1960s and early 1970s, RAND initiated a project called "RAND Project 1476, Analysis of Systems for Air Force Education and Training." This effort involved the development of a resource and cost model, a media model, and an airmen technical training flow model. Allison reports that the resource and cost model was completed and tested against costs developed using other methodologies

---

[2]H. E. Boren, Jr., *The Pilot Training Study: A User's Guide to the Advanced Pilot Training Computer Cost Model (APT)*, RAND Corporation, Santa Monica, CA, RM-6087-PR, Dec 1969, p. 1.

[3]This work was for internal RAND use only, Apr 1970. It is listed only as an example of some of the work that was occurring at the time.

[4]A. Hammond, *Mathematical Models in Education and Training*, RAND Corporation, Santa Monica, CA, RM-6357-PR, Sep 1970, p. iii.

employed by the Air Force.[5] The example in the report shows model estimates within one percent of the Air Force cost methodology estimates. Allison notes that due to the large number of inputs and the difficulty in getting data, it was not feasible to examine a large number of resident courses. The report includes a detailed description of the model and code.

The airmen technical training flow model was completed and renamed the Technical Training Requirements Model (TTRM). Kennedy describes the original model in a working note,[6] and Berman developed a user's guide.[7] According to Berman, the model was successfully run with real data.[8]

Development of the media model followed the previous two models. In February 1971, Bretz described and definitively explained communication and instructional media.[9] In June 1972, Samaniego developed "preliminary plans for a student flow model," with an emphasis on the impact of media selection in the training process.[10] Samaniego describes the basic elements and relevant characteristics of the proposed model.

Despite a continuing need for good analytical tools for training, none of these models is in use today.

In a nonmodeling research effort that occurred more recently, Robbert et al. studied the responsiveness of the training pipeline from the viewpoint of changing course requirements. Specifically, they looked at possible changes in the U&TW process that would

---

[5]S. L. Allison, "A Computer Model for Estimating Resources and Costs of an Air Force Resident Technical Training Course," RAND Corporation, Santa Monica, CA, working note, Oct 1970.

[6]P. J. Kennedy, "A Model for Estimating Technical Training Requirements," RAND Corporation, Santa Monica, CA, working note, Sep 1971.

[7]M. B. Berman, "User's Guide to the Technical Training Requirements Model," RAND Corporation, Santa Monica, CA, working note, Oct 1971.

[8]Ibid., p. iii.

[9]R. Bretz, *The Selection of Appropriate Communication Media for Instruction: A Guide for Designers of Air Force Technical Training Programs*, RAND Corporation, Santa Monica, CA, R-601, Feb 1971.

[10]M. G. Samaniego, "Preliminary Plans for a Student Flow Model," RAND Corporation, Santa Monica, CA, working note, Jun 1972, pp. 9–11.

improve its effectiveness.[11] The report was a stimulus in changing the responsibility for U&TW leadership to a dual-chaired process (AETC/DOO and the AFCFM).

Previous Air Force Human Resource Laboratory Research

Most of the Human Resource Laboratory work in this area over the past ten to twenty years has revolved around the development of the Training Decisions System (TDS) and the newer PC version, the Training Impact Decision System (TIDES). In a 1992 report, Keric et al. gave an overview of emerging technologies being used to develop a model that covers the entire career of an individual. The model simulates the flow of airmen based on a utilization and training pattern. Using information about the tasks airmen perform, the flow of personnel through the assignment system, and the training resources available, it determines capacities and cost-effective training options.[12] The model did not appear to be complete when we were writing this report, and the authors state that "developing a model of the (utilization and training) pattern itself requires a substantial data collection effort."[13]

In August 1995, Gosc et al. published an operational guide to the TIDES model. TIDES reduced the 26-step TDS process to 13 steps. Its inputs include various sections of the Career Field Education and Training Plan (CFETP), the Specialty Training Standard (STS), course control documents (such as training plans, plans of instruction [POIs], and lesson plans), and data from the descriptions of formal training courses. The model generates reports for utilization and training patterns, training quantities, resource requirements, and cost estimation, as well as an optimization summary.[14]

---

[11]A. Robbert et al., "Determining Course Content for Air Force Enlisted Initial Skill Training: How Can the System Be Made More Responsive?" RAND Corporation, Santa Monica, CA, internal draft, Jul 2001, pp. 35–36.

[12]B. O. Keric et al., *Introduction to Training Decisions Modeling Technology: The Training Decisions System*, Armstrong Laboratory, Brooks AFB, TX, AL-TP-1192-0014, DTIC AD-A249 862, Apr 1992, p. vi.

[13]Ibid., p. 15.

[14]R. L. Gosc et al., *Training Impact Decision System for Air Force Career Fields: TIDES Operational Guide*, Armstrong Laboratory, Brooks AFB, TX, AL/HR-TP-1995-0024, DTIC AD-A303 685, Aug 1995, pp. 11–15.

In December 1995, Mitchell et al. documented the continuing research and application of TIDES. TIDES data collection requires that a group of subject matter experts (SMEs) within each Air Force occupational specialty identify the jobs and training courses for the specialty. In addition, they must identify the individual tasks.[15] The SMEs validate the jobs, skills, and experience required to perform the job, the experience required to hold the job, and the number of years the job is to be performed. Cost information is gathered at each training site; it must cover both OJT and formal training. The report states that the authors used the model on several Air Force specialties.[16]

While TIDES appears to offer the ability to produce cost and capacity estimates for training pipeline problems, the required level of data is too great and the detail level of the model is too complex for AETC's quick-turnaround planning questions. Apparently, the Air Force is not using TIDES today, primarily because such extensive data are needed for a proper TIDES analysis. METRICA maintains TIDES and has used it for various individual analyses over the years, so its use is evidently feasible in some situations.

Official Air Force Documents

AFI 36-2201 is a six-volume set describing overall training policy. AFI 36-2201, Volume 1, requires formal schools to report unresolved constrained courses to AF/DPLT prior to the TPR conference.[17] HQ AETC/DOO must develop a standardized technical training surge template.[18] The FY2002 AETC Performance Plan states AETC's mission as "to sustain the combat capability of America's Air Force."

---

[15]B. M. Perrin et al. estimate that 500 to 1,200 behavioral statements are required for each occupation. Perrin et al., *Methods for Clustering Occupational Tasks to Support Training Decision-Making*, Armstrong Laboratory, Brooks AFB, TX, AL/HR-TP-1996-0013, DTIC AD-A311 368, Jun 1996, p. 2.

[16]J. L. Mitchell et al., *Research and Development of the Training Impact Decision System (TIDES): A Report and Annotated Bibliography*, Armstrong Laboratory, Brooks AFB, TX, AL/HR-TP-1995-0038, DTIC 1196-1106 153, Dec 1995, pp. 4–5.

[17]HQ USAF/DPDT, AFI 36-2201, *Vol 1, Training Development, Delivery, and Evaluation*, 1 Oct 2002, para. 3.7.3.

[18]Ibid., para. 3.8.2.

Business Literature

In the current business literature, much is discussed about enterprise resource planning (ERP) systems to support corporate decisionmaking through information technology. ERP is a recent catch phrase that has many implications. The primary goal of ERP is to attempt "to integrate all departments and functions across a company onto a single computer system that can serve all those different departments' particular needs."[19]

While this may sound attractive, "companies that install ERP do not have an easy time of it."[20] One of the big problems is cost. Meta Group has studied the total cost of ownership (TCO) across several companies. As it happens, the average TCO is not representative, because the organizations in the study are of such different sizes. However, on a per capita basis, by dividing TCO by the number of employees, Meta Group found the TCO per "heads-down" user across firms is $53,320.[21] To reemphasize, that is $53,000 per employee. ERP programs tend to overlook hidden costs in training, integration and testing, data conversion, data analysis, consultants, ongoing implementation teams, and waiting for the return on investment (ROI).[22]

The Air Force's recent MILMOD conversion is an ERP-type effort, as is the TTMS implementation. In both cases, the programs have struggled through many growing pains. The costs of both efforts have increased and are a good indication of the types of issues a new cost and capacity system would face if implemented in AETC. ERPs are very expensive efforts. Furthermore, as briefly discussed in Chapters Two and Three, timeliness needs for data in AETC's training management are an issue. Truly real-time data systems are very expensive. We do not believe a minute-by-minute system is cost-

---

[19]C. Koch, D. Slater, and E. Baatz, "The ABCs of ERP," *CIO Magazine*, http://www.cio.com/research/erp/edit/122299_erp.html, p. 1.

[20]Ibid., p. 2.

[21]Ibid., p. 4.

[22]Ibid., pp. 4–6.

effective for the types of decisions AETC makes based on its cost and capacity data. We do not feel the cost of implementing a real-time ERP system is worth the gain to AETC.

Literature on Organizational Design and Management Structure

In drawing our conclusions on the management structure of training (see Chapter Two), we consulted the literature on organizational design and management structure. Mainly theoretical in nature, that research supports our categorization of the training management structure. Organizations have five basic parts: three layers of management buffered by support staff and technostructure. The three layers of management are the strategic apex, the middle line, and the operating core. The strategic apex, which we call strategic training management, includes people with overall responsibility for the organization and global concern. The strategic apex ensures that the organization serves its mission, via direct supervision and resource allocation, interaction with the environment and outside stakeholders, and development of the organization's strategy. The middle line, analogous to our direct training management, provides coordination between the strategic apex and the operating core by collecting and disseminating information, intervening in the flow of decisions, and managing boundaries—effectively serving as CEOs of operating core organizations. The operating core, our direct training unit, has four major functions: secure inputs for production, transform inputs to outputs, distribute outputs, and provide direct support to input, transformation, and output functions. We consider the corporate Air Force, our fourth functional level, an additional strategic apex level, necessary to seamlessly incorporate a large organization like AETC into the rest of the Air Force. In addition to the management levels, there are support staff, who work outside the operating flow and include specialized support units. Examples within AETC are TRSS and the base infrastructure. Technostructure includes analysts who serve the organization by striving to make the operating flow more effective and efficient. This is roughly equivalent to the supporting offices within AETC management, which seek to improve the training organization.[23]

An aligned organizational structure enables three processes critical to modern organizations: decisionmaking, innovation, and information acquisition and distribution.[24] Our analysis focused on the importance of data collection and flow in support of informed decisionmaking. In the absence of an aligned and efficient training organization, information acquisition and distribution will be incomplete and thus will hamper effective decisionmaking. This supports our emphasis on aligning organizations in an effort to simplify data flows. A well-designed hierarchy (1) adds real value to work as it moves throughout the organization, (2) identifies accountability at each stage of the value-adding process, (3) places people with the necessary competence at each organizational level, and (4) builds consensus and acceptance of the managerial structure that achieves these ends.[25] Each additional managerial level is accountable not only for the work of subordinate levels, but also for adding value to that work.[26] Managerial levels that only pass information and act as go-betweens are repetitive and add unneeded complexity to a hierarchy.

CASE STUDIES

In the preceding chapters, we argue for a flatter AETC organizational structure in the interests of better defining decisionmaking roles and simplifying the data flows to decisionmakers. Much of the rationale for the current, five-level command structure (AETC/NAF/wing/group/squadron) at AETC is that all Air Force commands are to be defined like warfighting commands and that the span of control is to be limited at each level. Specifically, the argument for span of

---

[23]Henry Mintzberg, *The Structuring of Organizations*, Prentice-Hall, Englewood Cliffs, NJ, 1979.

[24]George P. Huber, "The Nature and Design of Post-Industrial Organizations," *Management Science*, 30(8), 1984.

[25]Elliott Jaques, "In Praise of Hierarchy," *Harvard Business Review*, 68(1), Jan/Feb 1990, pp. 127–134.

[26]Ibid.

control limits is that it is infeasible to have all TRGs responsible for technical training report directly to the 2AF/CC or equivalent.

The counterargument is based on the functional analysis of training presented in Chapter Two. AETC's primary technical training activities are accomplished in the TRSs. TRGs provide a location for administration and management functions (such as record keeping, maintenance, and technology development) that are subject to economies of scale. Above that are requirements for strategic training management. A side function relates to base operations, which need to be in the purview of the wing commander but do not need to be in the functional chain of command.

We now turn to an examination of training organizations other than AETC in order to gain insight into other structural options. First we look at Air Force Materiel Command (AFMC), which is not organized like other MAJCOMs. We then consider the military training organizations within the Army and the Navy. Finally, we examine training organizations in industry, drawing parallels where appropriate.

Air Force Materiel Command

Clearly, AFMC does not perform the same function as AETC, but it is a MAJCOM whose organizational structure does not conform to that of other MAJCOMs.

Figure A.1 displays an organization chart for AFMC. Note that this structure is much flatter than AETC's. All of the air logistics centers and numerous special technical organizations report directly to the AFMC/CC. Most of these organizations are stand-alone, have well-defined activities, and need little direct supervision (as is true for AETC's training organizations). It is also interesting that the primary function is considered to be the center, not the base. Base support services, when required, are provided by an air base wing reporting to the center commander.

The AFMC structure is similar to the flatter organization we propose for AETC. It appears that AFMC has been able to manage any problems that could potentially result from such a broad span of control. This kind of structure may also enable AETC to reduce organizational overlap and the need to pass information between so many levels.

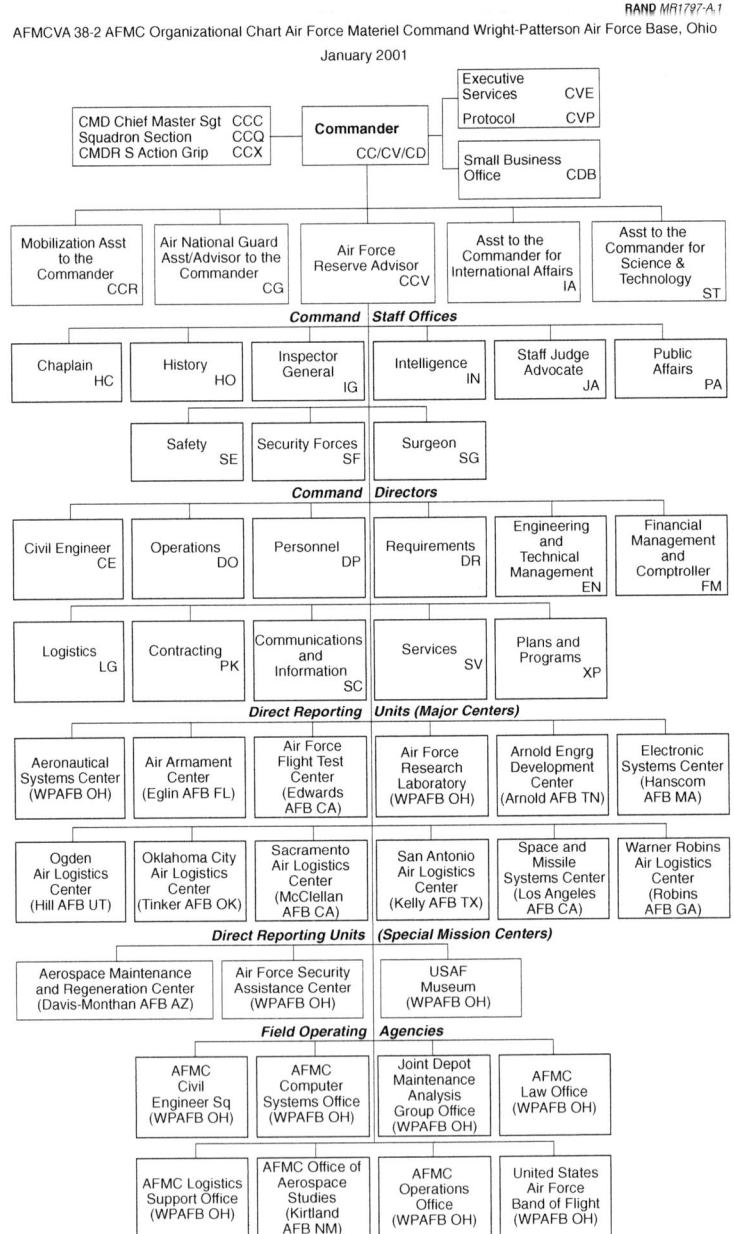

RAND MR1797-A.1

Figure A.1—AFMC Organizational Structure

Training in the Other Services

The Air Force is, of course, not the only service with an extensive formal training program. The other services also provide initial training to entering enlisted personnel. Both the Army and the Navy have an organization that corresponds to AETC, although the structures of these organizations differ from that of AETC as well as from each other. We begin by examining the Army's system and then move to the Navy's.

Army. The Army organizes training under the Training and Doctrine Command (TRADOC), which is responsible for both the development of doctrine and the training to implement doctrine in various functional areas. These include such warfighting disciplines as armor and infantry (including their associated maintenance functions), as well as such support functions as signal corps and quartermaster.

Figure A.2 shows a template of the current training and doctrine organizational structure. Underneath HQ TRADOC are training centers grouped by primary functional area (e.g., the Armor Center at

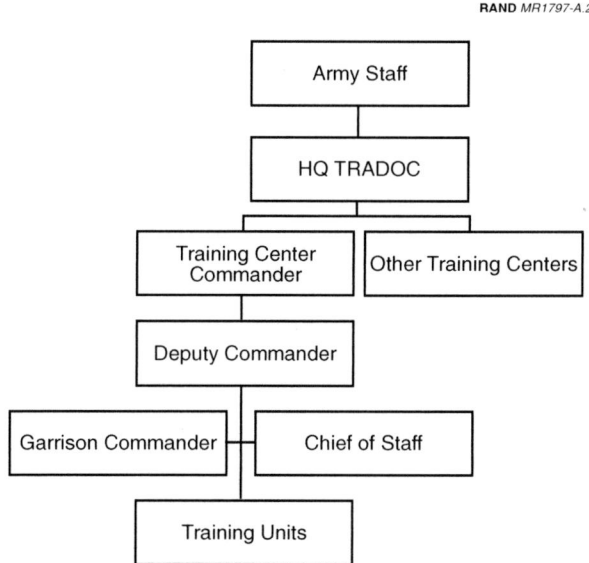

RAND *MR1797-A.2*

Figure A.2—Idealized Army Training Structure

Fort Knox, the Signal Center at Fort Gordon, and the Quartermaster Center and School at Fort Lee). Unlike the Air Force, the Army organizes its training activities for both the enlisted force and officers strictly along these warfighting functional areas. In the template organization, the center commander has three primary subordinate officers: the deputy commander, who is in charge of the training regiment(s); the chief of staff, who is responsible for management and administrative functions; and the garrison commander, who supervises BOS functions.

Figure A.3 illustrates the high-level breakout of the organizational functions at the Signal Center and Fort Gordon (SC&FG). Although each training center is organized slightly differently in detail, SC&FG serves as an example of how the training functions are generally organized within the Army training centers. The regimental directorate of training (RDOT) performs the bulk of the training management functions, with support from some offices under the chief of staff, such as multimedia services under the director of plans, training, and mobilization. Direct training is performed primarily within the training divisions, with support from the regimental training support division (similar to the Air Force's TRSS). Strategic training management is split between the center commander (to handle management within the functional area) and HQ TRADOC (to handle management across functional areas). Support for strategic training management within a center is primarily split between the chief of staff and the garrison commander. In particular, the director of resource management performs policy, analysis, and budgeting functions. Although training facility and BOS requirements and execution of training infrastructure fall under the garrison commander, the infrastructure budget is compiled along with the training mission budget under the director of resource management.[27]

The Army training mission is highly influenced by its organizational structure. The organizational levels are flatter than the Air Force's training organization levels and are more directly associated with the training management levels we identified. Moreover, the focus of the centers is based on wartime fighting functions.

_____

[27]Under the chief of staff are also the doctrinal functional organizations such as the chief of signals, TRADOC system managers, and the director of combat development.

RAND *MR1797-A.3*

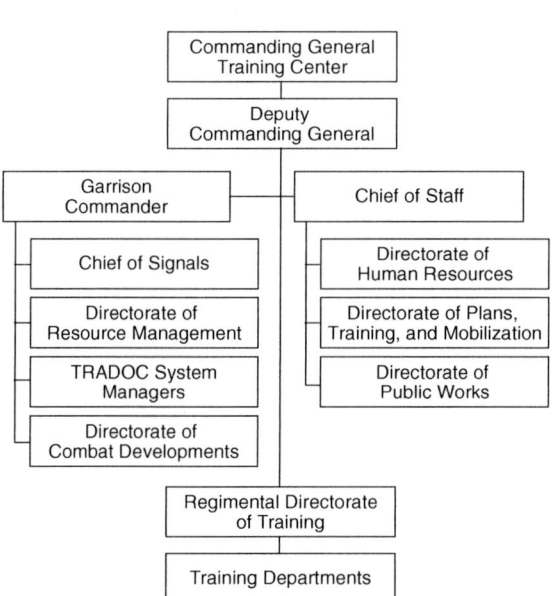

**Figure A.3—Simplified Functional Organizational Chart
for Army's Signal Center and Fort Gordon**

Imbedding both doctrine and training in the organization has both positive and negative effects. The positive effect is that training across all aspects (including enlisted and officer training) of the functional area is more easily coordinated, and advocates for the functional missions are clearly defined. The negative effect, however, is stovepiping and the increased difficulty of providing training for linkages across functional areas and establishing priorities among them. Collocating direct training, training management, and strategic training management within a center would make communication, coordination, and management tasks—including data collection and distribution—easier. However, it would complicate strategic training management by splitting it up among multiple centers.

Navy. It is a very interesting time to be looking at the Navy's organizational training structure. The Navy is currently involved in a large effort to shift its focus from the old "brick-and-mortar" initial-training schoolhouse concept to a career-long approach to training and education. This reorganization and conceptual shift have been dubbed Task Force EXCEL. For the short term, the Task Force EXCEL offices are serving as the Navy's catalyst for the training revolution; eventually they will be reabsorbed into the Navy's traditional training organization, CNET. The Chief of Naval Operations tagged this revolution in training his top priority in FY2002.

The revolution in training was inspired by the Executive Review of Naval Training (ERNT) completed in August 2001, which was a complete look at the Navy's training and education over a sailor's career. The review was designed to examine Navy training and make substantive recommendations for improving and aligning organizations, incorporating new technologies into Navy training, exploiting opportunities available from the private sector, and developing a continuum of lifelong learning and personal and professional development for sailors. We are particularly interested in the Navy's first aim, that of examining its current training organization, and in its recommendations for improving and aligning that organization. Most of the work thus far within Task Force EXCEL has focused on incorporating new technologies within training and establishing the sailor continuum that outlines career development. We are unaware of any reorganization to date.

Currently, there is no central organization for training in the Navy. While CNET is certainly the Navy's largest provider of training, it is not the center for all training and does not make all strategic training management decisions. According to the ERNT: "No single organization is responsible for Navy training and education. For example, [CNET] is responsible for most of initial skills training, much of advanced skills training, but little of officer education." Because there is no one central training organization, there is no overarching training strategy to guide Navy training. To make this point further, the ERNT lists 63 organizations that can impose training requirements, 38 that perform training management functions, and 39 that coordinate training exercises. There is no mechanism for coordination among these organizations, especially in imposing training requirements. This means that two electrical technicians, one in the Atlantic fleet and one in the Pacific fleet, can face different training requirements. Having one central training organization provides a training authority that can coordinate cooperation among training organizations.

Consequently, CNET is not the only organization receiving training dollars. Currently, 11 resource sponsors at the HQ U.S. Navy Staff (OPNAV) level (equivalent to Air Force's Air Staff) send money to 13 claimants who provide training. CNET is the largest of the 13 claimants, but training dollars also go to organizations such as Naval Reserve, Naval Personnel, the Atlantic and Pacific fleets, and Naval Air Systems Command. The funding of Navy training and education is fragmented and decentralized at the OPNAV level into a complex, platform-centric environment. Individual OPNAV departments manage the training dollars for their specific programs.

There is no single advocate for training among the OPNAV departments. In addition to the fragmented funding sources at the OPNAV level, no central organization owns training. This means there is no central office to gather information or advocate for training. Currently, Training and Education is a division directorate, N79, under the Warfare Requirement and Programs department, N7. The ERNT recommended that there be a department at the OPNAV level to deal exclusively with training issues and to consolidate the current fragmented funding and oversight responsibilities within one office. The Air Force might also consider the benefits of having an office at the Air Staff that advocates exclusively for training.

Within CNET, training is organized according to the associated Navy platform or discipline. As an example, a chief of Naval Air Training is in charge of all training within Navy aviation. The Naval Submarine School in Groton, CT, is the single provider for basic enlisted submarine school and associated "A" schools. These functional schools provide all platform-specific training as well as initial skills training for platform-specific ratings. The majority of initial skills training that is not platform-specific falls under the Services School Command at Great Lakes. This is where both basic training and most "A" school training are completed for the Navy. The Services School Command is split into the Engineering Systems School and the Combat Systems School. The former trains ratings who maintain the ship's power plant; the latter is divided into three divisions: advanced electronics, gunnery school, and seamanship. The chief of

Naval Air Training, commanders of each platform-specific school, and the Services School Command report directly to CNET.

Unlike AETC, the Navy has different organizations controlling recruiting and initial recruit training. The Navy's Recruiting Command focuses on recruiting men and women for enlisted, officer candidate, and officer active duty status in the regular and reserve components of the Navy. It primarily focuses on meeting a recruiting quota for accessions. The Recruit Training Command, under CNET, is responsible for conducting boot camp and transitioning Navy recruits into sailors. Each organization focuses on completing its individual mission within its limited resources.

At times, the two organization's missions are in competition, making them more difficult to accomplish. For example, in an effort to meet recruiting goals, the Recruiting Command may recruit lower-quality personnel, making it harder for the Recruit Training Command to transform them into competent sailors in jobs across the spectrum of ratings. The ERNT discussed this problem and suggested that the objectives be aligned by making the two organizations into one with a consolidated mission: transforming civilian recruits into trained sailors ready for initial skills training or to join the fleet. Then tradeoffs between recruit quantity and quality could be made inside one organization, ostensibly leading to more-efficient solutions. It is unclear whether the Navy will move in this direction, but the Marine Corps and Air Force are successful examples of aligned organizations in recruiting and recruit training.[28]

One difference between the Navy and Air Force that complicates initial skills training for the Navy but adds pipeline flexibility is that the Navy does not assign all sailors their ratings (similar to an AFSC) by the completion of boot camp. Some sailors who are undecided on their desired rating go directly to the fleet and work as undesignated seamen without initial skills training. These sailors spend time at sea learning about different jobs in the Navy and then "strike" for a rating. This does not guarantee that they will receive their chosen rating, but once classified with a rating, they proceed to initial skills training, their "A" and "C" schools. This model has worked well for

---

[28]While recruiting is not discussed in any of the other case studies, the experience of the Navy supports the current Air Force structure of aligning education and recruiting.

the Navy because some jobs in the fleet do not require training in a specific rating and can be performed by undesignated sailors. This tends to give the Navy more flexibility in filling job categories, because there is a group of sailors trained through boot camp who could be called directly to initial training for a particular rating in times of shortage. For example, when the Air Force instantly needed more security forces after 9/11, it might have been helpful to have a set of unclassified airmen who had completed basic training and could be immediately channeled into that AFSC without drastically disrupting the AFSC assignment distributions of airmen in the pipeline. However, it is not clear how many Air Force jobs undesignated airmen could fill.

The Navy's equivalent to the Air Force's U&TW process is the Navy Training Requirements Review (NTRR), which is held once every three years. Its purpose is to establish a cyclical review of shore-based training; its focus is, similarly, to ensure that shore-based training satisfies fleet requirements. The participating members are resource and program sponsors, representatives from the fleet, Bureau of Naval Personnel (BUPERS), Navy Manpower Analysis Center (NAVMAC), and training agents (i.e., CINCPACFLT, CNET, etc.). Navy training plans are implemented to correct course deficiencies identified in the review and to realign shore training with fleet requirements. A large effort goes into ensuring that classroom training stays abreast of new technologies aboard ships.

Industry

Many U.S. industries have developed extensive in-house training organizations. This was particularly the case in the 1990s as they endeavored to ensure that their employees' skills stayed current in a rapidly changing world.

Although many of the training organizations that companies have developed do not focus on entry-level training, we selected several companies from *Training* magazine's Top 100 list for 2002 and contacted their training organizations for unstructured telephone or in-person interviews regarding their strategic management structure. The following sections provide a thumbnail sketch of the findings from each interview, including the organization's size and scope, the company's organization of its strategic management of training, information about training assessment, and interviewee comments on forthcoming training challenges.

Automobile Manufacturing University. Automobile Manufacturing University (AMU) is the professional training organization for a large North American automaker. It is five years old and teaches 1,500 courses providing 190,000 hours of training annually to white-collar workers. It does not train production workers, and it does not provide initial training. We interviewed the manager of AMU.

To the students, AMU's structure is a set of colleges that are aligned with the company's functional processes: communications, engineering, quality, etc. Internal decisionmaking is divided into two pieces:

- Curriculum decisions ultimately rest with global process leaders, senior executives in the process areas. They are advised by "deans," functional area experts responsible for assessing job demands and current training needs, possibly augmented by one or two other staff members.

- Execution and strategic direction are provided by the AMU organization. This includes enrollment recording, course catalogs, instruction contracting, and new media development.

AMU staff are divided roughly equally between the two areas, for a total of about 100. AMU has no staff instructors: it outsources instruction, course development, and course upgrades (directed by the deans). In theory, each college could obtain courses outside AMU, but the courses have been tailored to the company's needs and practices.

Decisions on resource allocation are made between the colleges and AMU itself. The colleges propose curricula; AMU prices the courses for cost recovery plus overhead (and periodically benchmarks against outside providers). Ultimately, the using organizations have to vote by paying for course attendance.

Assessment is done first with students, deans, and instructors. In-class exams are also used for most courses. Managers rate the most important classes by observing the performance of trained employees. They use a five-point scale and then simplify it to a color scale.

Satisfactory, green, is 4.5 to 5. Any ratings under 4 are red, meaning that the course will be reconsidered.

The AMU manager believes that the biggest challenge confronting his training organization is satellite and digital course delivery. Two years ago, 90 percent of the training was standard classroom format. Today, 28 percent is classroom via satellite and 14 percent is e-training. The key is to decide what can be digitized to cut training costs. The savings are not primarily in course development, but in employee time. The manager noted that for some types of training, physical attendance is better.

Motorola University. Motorola University (MU) is the corporate training arm of Motorola Corp. It has been active for over ten years and has gone through several different organizational structures: centralized, decentralized, different training models, etc. At present, Motorola has many different business types, so most training has been delegated to individual business units. The responsible organization in each unit is a BST (business support team), a multifunctional team of training specialists whose leader is a senior human resources staff member. The BSTs assess the needs of the business unit, set priorities, and choose training modalities and providers.

Motorola provides both entry-level and ongoing training. It expects each of its 110,000 employees to spend five days per year in training. Each person has an individual development plan, reviewed quarterly, to help improve competence on the job. Training is available in technical areas as well as in communications and leadership.

MU itself focuses on only two types of training: leadership development and quality.[29] These two areas are considered core for the company, and the CEO wants all employees to have a common language to discuss these issues. In addition to these two foci, MU is developing expertise in e-training that it plans to market to the business units.

Entry-level training is starting to move to virtual reality training in place of OJT. Our contact said their early experience shows that in

---

[29]Motorola pioneered the six-sigma concept of quality development and has taught these methods to thousands of students both inside and outside the company.

some cases the time to train has been reduced from six months to three weeks because students have more freedom to experiment (in a safe environment) and need not wait to see a complex problem in the real world.

Northwest Airlines. Northwest Airlines (NW) has decentralized its training to individual organizations. We interviewed the Director of Technical Training, who works for the Vice President of Technical Operations, who is primarily responsible for all areas of aircraft and equipment maintenance. Flying Operations runs pilot training, and similar arrangements exist for the baggage handling and gate/check-in areas. The Technical Training (TT) staff has about 100 professionals, of which 70 are instructors (the others are instructional system designers, Web maintainers, and clerical support). They are spread over four training hubs and provide 280,000 student hours per year to 8,000 internal and 13,000 external students.

Unlike the other organizations we interviewed, NW TT is a corporate function and is provided free to the functional areas. In this sense, the NW experience is close to that of the Air Force. TT is responsible for all of the strategic planning for training, including budgeting, capacity planning, curriculum changes, and improvement. However, it must also justify its performance and existence to its customers and the corporation. For this reason, NW TT pays particular attention to assessment and metrics.

The current TT director brought to NW and expanded an approach he had developed while in a similar position at FedEx. Called the Training Quality Index, it consists of five types of metrics:

1. Effectiveness: How good is training? NW TT uses pre-post tests, which may be knowledge- or performance-based as appropriate.

2. Quantity: Is there enough training? This includes capacity, seat utilization,[30] student hours, and operational requirements for training (elicited from the operational areas as part of a business plan).

---

[30]NW TT aims for 70 to 85 percent seat utilization to maintain flexibility. Our interviewee argued that if full utilization is the metric, then there is an overwhelming incentive to undertrain.

3. Perception: What do managers and workers think of their training? This is critical, because poor perceptions need to be corrected by data if they are wrong or acted on if they are justified.

4. Cost: Cost per student and budget variance.

5. Operational impact: This is *very* difficult to determine, because so many other things can affect operational performance. The focus is on specific problem areas where training has had a demonstrable effect.

The training inputs use a Web-based data entry form that is filled out by instructors (and designed to minimize their work). Inputs include the professional use of the instructors. Cost data are collected from financial systems, and perception inputs come from surveys and personal contacts. The result is a monthly paper report that includes information at all levels of aggregation from systemwide down to individual instructors.

Boeing.   Boeing's St. Louis factory produces a number of different military systems. As with other organizations, training at the facility has been split into a number of different areas, with TT responsible for the initial and continuing education of production workers and manufacturing support personnel. The position reports to the Director of Manufacturing Support; actual production is controlled by program directors responsible for the manufacturing of individual systems (e.g., the F-15). TT has 27 full-time employees (instructors and course developers) and 25 part-time instructors drawn from the shop floor. Their target population is 1,700 production workers, to whom they provide 50,000 hours of training per year.

TT provides and develops virtually all courses that deal specifically with technical issues (some training on other issues such as computer security may be provided by the home office). The courses are planned based on upcoming program needs and are presented to the program directors, who must decide how much training their programs can afford. The courses and plans are then negotiated between TT and the programs, although higher management may intervene if the programs cannot fund some critical training needs.

Facilities and training equipment are also the responsibility of the training manager, who makes proposals for new equipment and new or renovated space to central boards (one for capital assets and one that manages all facilities). After the proposals are gathered across departments and within the facility and are prioritized, resources are allocated.

METHODOLOGICAL CONSIDERATIONS

We have identified three reasons why the Air Force needs initial skills cost-to-train data for each AFSC. First and foremost, AETC wants a dollar amount that represents the cost to train for a specific AFSC, a figure that can be used for justification in the POM process. While AETC can compute the costs involved in training, the general consensus from our interviews is that these costs are incomplete (too low). A more robust and accurate capability would help defend AETC's budget against cuts and make it possible to show production increases for a given funding increase. It is very difficult to show the direct effect of training on an end measure such as operational readiness, but it is possible to see the indirect effect of training through measures such as manning levels. With a link between training budgets and training production through a cost-to-train figure, AETC can make a first step toward relating its budget to readiness, through the budget's effect on manning.

Second, the Air Force uses these cost figures to charge foreign governments who send personnel to Air Force training programs.[1] It is unclear whether the charges to foreign governments currently cover all the costs associated with their training.

Third, with accurate cost-to-train figures, the Air Force could make TPR tradeoff decisions that maximize readiness by producing efficient quantities in each AFSC. This idea can be compared to the

---

[1]AETC does have the capability to compute cost-to-train for foreign exchange students. However, it is a very time-consuming process and is not a complete accounting of all costs.

economic concept of a perfectly competitive market where the pro-
ducer maximizes his surplus by producing a quantity where price
is equal to marginal cost. The optimal situation is one in which an
efficient transfer market is established between AETC and the
MAJCOMs wherein efficient levels of airmen are produced in each
AFSC, yielding the highest return for a given training investment. Is
this market simply theoretical, or would it be possible if accurate
cost-to-train data were available?

AVERAGE VERSUS MARGINAL COST PRICING

When computing cost-to-train, one needs a dollar value that cap-
tures all of the cost to the Air Force of producing each airman in a
given AFSC. Typically, there are two types of economic costs that can
identify this cost per airman: average and marginal costs. Average
cost denotes the total costs of providing the training divided by the
total number being trained. This type of computation distributes
large fixed costs over all airmen produced. Fixed costs (e.g., the large
capital costs for dorms, dining halls, and classroom space) cannot be
varied in the short run. In the long run, expanding capacity can
increase these production inputs, but they are fixed in the current
production year.

The cost calculation for marginal cost asks, What is the cost to train
one additional airman? This view assumes that fixed costs have already been paid, so marginal costs reflect only the incremental cost
of producing one more or one fewer airman. In economics, marginal
costs are useful in free-market firms' production decisions. A firm
will operate at maximum efficiency when it produces at a level where
its marginal cost of production equals the market price. A firm that
bases it production decisions on average cost will tend to under-
produce because average costs generally are higher than marginal
costs over the relevant production range.

An average cost-to-train for each AFSC is useful for accounting purposes and may also be useful in charging foreign governments. But it
is not useful for production decisions, because it may not accurately
reflect the correct costs.

We illustrate this concept using an oversimplified example that reaches a different production decision depending upon whether one uses marginal-cost or average-cost pricing.

Suppose a training organization has ten more dollars it can spend and wants to maximize the number of additional trained airmen produced. Assume that marginal costs are constant and that mechanics and weathermen are of the same value. Table B.1 shows what the production decision looks like.

Based upon average costs, weathermen appear to be cheaper to produce. But the available $10 will yield only an additional five weathermen ($10 divided by marginal cost of $2) compared to ten mechanics. Mechanics look more expensive than weathermen based on average cost—due to the high fixed costs of the mechanics' training devices. But once those fixed costs are covered, the marginal costs reveal that it is cheaper to produce additional mechanics—ten mechanics are produced for $10.

Although oversimplified, this example helps to show a basic point. If a cost-to-train value is needed for accounting purposes or for charging foreign governments, the average cost (including fixed costs) yields the right number. If, however, a cost for setting production levels is desired, it is important to separate fixed costs from the production decision and concentrate on the marginal production cost. When doing so, one must have a good sense of production capacity to avoid setting a requirement to train more students than can possibly be trained. We believe that the Air Force cannot make this production tradeoff between AFSCs, because it does not have these types of data. This information would be useful to have when making tradeoff decisions between AFSCs—for example, during the TPR conference.

Table B.1

Simple Example of Costing an AFSC

	Fixed Cost ($)	Marginal Cost ($)	Cost to Produce 100 ($)	Average Cost ($)	Number Produced (for $10)
Weatherman	100	2	300	3	5
Mechanic	1,000	1	1,100	11	10

ACCURATELY ATTRIBUTING COSTS TO SPENDING ORGANIZATIONS

How does AETC go about converting raw data on costs into useful average and marginal cost numbers? One could calculate average cost simply by adding up all the costs and dividing by the quantity produced. Unfortunately, this simple approach will not work in this case, for several reasons.

We now look specifically at dividing overhead costs and the methodology for computing capital costs.

Dividing Overhead Costs

Costs are not subdivided by AFSC when budgeted. Some costs are not directly associated with a particular AFSC because they are executed at the group level or higher. As an example, should the costs for staffing and running the training support squadron (TRSS) be divided among the several AFSCs the TRSS supports? Costs often cannot be tagged to a particular AFSC and therefore must be divided among several. This action requires a specific business rule.

In the past, the best AETC could do was calculate costs per student using O&M costs and dividing by student throughput at each base, sometimes with adjustment factors. The cost of training was the same for all AFSCs at a given base because only base-level costs were known. AETC/DORB has done some work in estimating costs at a lower level, which (we hope) will be reduced to individual AFSCs. DORB's work divided higher-level costs among flights based upon number of training days. This issue was also identified as an action item for AETC/XP at the 27 June 2002 Technical Training Summit.

Methodology for Computing Capital Costs

The second problem with converting raw cost numbers into a useful decisionmaking support tool is the issue of distributing fixed capital costs over the life of an investment. Raw cost data on capital assets are available, but currently there is no methodology to distribute these costs over the life of the asset or, further, to individual students. Part of the difficulty in establishing a methodology is dealing with the uncertainty of facility lifespan, facility upgrades, and future student throughput.

The projected, useful lifespan of MILCON projects has changed year to year depending upon MILCON budgets. *Facility recapitalization rate* is a term used to identify the number of years under current and projected MILCON spending needed to replace the current capital stock. However, the idea of facility recapitalization rate is just a macro-level planning tool; it is not the solution to capital replacement. Air Force needs and facility requirements still drive the budgeting for capital replacement, making the replacement schedule even more uncertain. The primary measure for the lifespan of capital assets has been this facility recapitalization rate, which oscillates wildly and depends heavily upon current and projected MILCON spending. Without a good estimate for the lifespan of capital assets, it is difficult to depreciate the value and attribute facility costs to a given year or to individual students.

Some preliminary ideas have been discussed for creating a methodology for distributing capital costs annually to help calculate yearly cost to train, but much more work is needed in this area. One idea recognizes that the Air Force could lease capital assets from private companies rather than build its own, thus converting the cost of capital into yearly leasing costs. Effectively, this transfers to the private sector the problem of the time value of money, allowing the Air Force to simplify its accounting to a yearly cost of capital. While leasing would simplify the yearly cost of capital calculations, the option may prove to be more expensive, because the leasing agent will demand a return on initial capital investment. This is similar to the housing privatizations already occurring at Air Force bases, converting the upfront MILCON cost for the government to an amortized leasing schedule. The government has reportedly realized savings through privatization of 5 to 1 over traditional MILCON spending.

Another idea is to create a revolving fund to pay for the replacement of some number of buildings every year on a predetermined basis. Once all the buildings were replaced, one would start over with the first building. If AETC replaced the same amount of capital assets every year, it would have a good number for the annual cost of capital replacement. This idea would be very difficult to operationalize, however. First, Congress appropriates MILCON budgets yearly, so

AETC could not set the replacement rotation unless these controls were released to it. Second, a replacement schedule may not be in the Air Force's best interest. Air Force needs and facility requirements are what currently drive the MILCON process, not a set replacement schedule. Third, a revolving fund would require capital replacement to be smoothed over time, which probably is not consistent with the fluctuations associated with defense budgeting. This idea is similar to the current facility recapitalization rate computation that provides the number of years it will take to replace the entire capital stock given current and forecasted MILCON spending.

And as a final idea, if AETC wants a cost-to-train figure purely for accounting purposes, a good estimate of the yearly cost of capital can be obtained by establishing a business rule that simply depreciates a facility's cost over the asset's useful lifespan. Even though the actual costs are not amortized over the lifespan of the facility (i.e., MILCON money is paid up front), the Air Force continues to reap value from the asset over time and can distribute these costs to all students over the asset's useful lifespan. This accounting practice does not represent the actual yearly cost of the asset but could give an accounting estimate for charging foreign governments and justifying POM budgets. This type of calculation would require assumptions about two uncertain variables at the time of construction: the facility lifespan and facility depreciation rates.

The Time Value of Money

Another challenging dimension for any governmental organization trying to determine accurate cost data is dealing with the limited time value of money. One example of AETC's struggle with this problem is in its TDY-to-school budget. AETC sends students TDY to technical schools that last 20 weeks or less. The money for sending airmen TDY to their technical schools is centrally controlled by AETC from its O&M budget. This TDY-to-school bill has been growing; it consumes about $80 million of TT's approximately $200 million O&M budget. This is the largest piece of AETC's O&M budget.

In looking for ways to trim these costs, AETC found it was spending $5.1 million per year at Maxwell Air Force Base to lodge airmen off base because of dorm shortages. AETC must continue to pay these yearly O&M costs because it cannot invest this money in a new building project. It is the wrong kind of money, and because AETC cannot borrow against future budgets, it cannot afford the full upfront costs of building a new dorm in a single year. In industry, a company will invest in capital when it can realize yearly cost savings amortized over future years by introducing new capital today. Current government financial management does not allow for these cost savings. Imagine a capital account that would allow AETC to build a new dorm today and pay for it over the next several years with money it will save from reduced lodging costs. This type of capital account would introduce a time value of money to the government's decision.

Recent nontraditional MILCON funding ideas—e.g., housing and utilities privatization and enhanced-use leasing—have grown the partnership with industry and yielded cost savings for the government. Partnerships with industry have become the new fad in Washington, opening the way to future initiatives that will allow a time value of money for government spending. Future RAND research will continue to explore partnerships and collaborations with industry as well as independent initiatives that introduce time value of money to help solve problems like the one AETC faced at Maxwell.

CONCLUSIONS

AETC would like to have ways to calculate cost-to-train figures for each AFSC.[2] As mentioned in our discussion of facility and BOS costs in Chapter Three, data collection and methodological barriers prohibit the creation of the needed business rules. AETC is working on more accurately attributing high-level costs to subordinate organizations in two ways:

1. Improving data tracking techniques so that costs are more accurately attributed to the spending organization.

2. Establishing a way to divide grouped costs into subcomponents based upon a utilization factor, such as number of training days.

_____

[2]AETC/FM's Cost Modeling System (COSMOD) provides some allocation to subcomponents based on a utilization factor. It does not include capital costs.

Even if a methodology were developed to solve the capital and de preciation aspect and calculate a yearly facility capital cost, the problem of distributing costs to particular AFSCs remains. Training facilities are maintained and managed at the wing and group level, and utilization rates by AFSC can vary throughout the year depending upon class size and start dates. Business rules for attributing higher-level costs to subordinate entities should apply here.

BIBLIOGRAPHY

Adams, J. R., M. Parsons, and R. Hutchins, *Fiscal Year 2001 Training Cost Factors Booklet*, HQ AETC/FM, Randolph AFB, TX, undated but approximately 2001.

AETC Strategic Plan, 2001 (updated on AETC Web site, http://www.aetc.randolph.af.mil).

Allison, S. L., "A Computer Model for Estimating Resources and Costs of an Air Force Resident Technical Training Course," RAND Corporation, Santa Monica, CA, working note, Oct 1970.

Berman, M. B., "User's Guide to the Technical Training Requirements Model," RAND Corporation, Santa Monica, CA, working note, Oct 1971.

Blodgett, M., "GamePlans," *CIO Magazine*, http://www.cio.com/archive/011598/game_content.htm, 15 Jan 1998.

Boren, H. E., Jr., *The Pilot Training Study: A User's Guide to the Advanced Pilot Training Computer Cost Model (APT)*, RAND Corporation, Santa Monica, CA, RM-6087-PR, Dec 1969.

Bretz, R., *The Selection of Appropriate Communication Media for Instruction: A Guide for Designers of Air Force Technical Training Programs*, RAND Corporation, Santa Monica, CA, R-601-PR, Feb 1971.

Coffey, M., L. Kelly, and M. Parks, *Enterprise Resource Planning*, http://www.geneseo.edu/~mpp2/erppaper.htm, 30 Oct 2000.

Daley, G. A., D. G. Levy, T. Kaganoff, C. II. Augustine, R. Benjamin, T. K. Bikson, S. M. Gates, and J. S. Moini, "A Strategic Governance Review for Multi-Organizational Systems of Education, Training, and Professional Development," RAND Corporation, Santa Monica, CA, internal draft, Mar 2002.

Ehemann, P., "Technical Training Manpower Standards," 81TRW/ MO, briefing slides, undated.

Gallegos, M., *MODIA: Vol. 4, The Resource Utilization Model*, RAND Corporation, Santa Monica, CA, R-1703-AF, Jul 1977.

Gosc, R. L., J. L. Mitchell, J. R. Knight, B. M. Stone, F. H. Reuter, A. M. Smith, T. M. Bennett, and W. Bennett, *Training Impact Decision System for Air Force Career Fields: TIDES Operational Guide*, Armstrong Laboratory, Brooks AFB, TX, AL/HR-TP-1995-0024, DTIC AD-A303 685, Aug 1995.

Grasping the Future: Comparing Scenarios to Other Techniques, ManyWorlds.com, http://www.manyworlds.com/1/content/ WhitePapers/CO4120214515558.pdf, 3 Sep 2001.

Grotevant, S. M., *Business Engineering and Process Redesign in Higher Education: Art or Science?* CAUSE 98 Seattle WA, http://www. educause.edu/ir/library/html/cnc9857/cnc9857.html, 8 Dec 1998.

Haley, M. J., "Preliminary Studies for the Airman Training Model," RAND Corporation, Santa Monica, CA, internal draft, Apr 1970.

Hammond, A., *Mathematical Models in Education and Training*, RAND Corporation, Santa Monica, CA, RM-6357-PR, Sep 1970.

Hess, R. W., and P. Kantar, "MODIA: The Cost Model," RAND Corporation, Santa Monica, CA, working note, Jun 1976.

HQ AETC/XP, *FY2002 AETC Performance Plan*, HQ AETC/XP, Randolph AFB, TX, Aug 2001.

HQ USAF/DPDT, AFI 36-2201, *Vol 1, Training Development, Delivery, and Evaluation*, http://afpubs.hq.mil, 1 Oct 2002.

HQ USAF/DPDT, AFI 36-2201, *Vol 2, Training Management*, http://afpubs.hq.mil, 17 Oct 2002.

HQ USAF/DPDT, AFI 36-2201, *Vol 3, On-the-Job Training Administration*, http://afpubs.hq.mil, 30 Sep 2002.

HQ USAF/DPDT, AFI 36-2201, *Vol 4, Managing Advanced Distributed Learning*, http://afpubs.hq.mil, 23 Oct 2002.

HQ USAF/DPDT, AFI 36-2201, *Vol 5, Career Field Education and Training*, http://afpubs.hq.mil, 27 Sep 2002.

HQ USAF/DPDT, AFI 36-2201, *Vol 6, Total Force Training and Education Review Process*, http://afpubs.hq.mil, 27 Sep 2002.

Huber, G. P., "The Nature and Design of Post-Industrial Organizations," *Management Science*, 30(8), 1984.

Jaques, Elliott, "In Praise of Hierarchy," *Harvard Business Review*, 68(1), Jan/Feb 1990, pp. 127–134.

Kennedy, P. J., "A Model for Estimating Technical Training Requirements," RAND Corporation, Santa Monica, CA, working note, Sep 1971.

Keric, B. O., T. A. Lamb, W. R. Bennett, and D. S. Vaughan, *Introduction to Training Decisions Modeling Technologies: The Training Decisions System*, Armstrong Laboratory, Brooks AFB, TX, AL-TP-1992-0014, DTIC AD249 862, Apr 1992.

Koch, C., D. Slater, and E. Baatz, "The ABCs of ERP," *CIO Magazine*, http://www.cio.com/research/erp/edit/122299_erp.html.

Malhotra, Y., *Role of Information Technology in Managing Organizational Change and Organizational Interdependence*, BRINT Institute, http://www.brint.com/papers/change, 1993.

Mintzberg, Henry, *The Structuring of Organizations*, Prentice-Hall, Englewood Cliffs, NJ, 1979.

Mitchell, J. L., W. Bennett, W. E. Wimpee, G. R. Grimes, B. M. Stone, and F. H. Rueter, *Research and Development of the Training Impact Decision System (TIDES): A Report and Annotated Bibliography*, Armstrong Laboratory, Brooks AFB, TX, AL/HR-TP-1995-0038, DTIC 1196-1106 153, Dec 1995.

Mooz, W. E., *Pilot Training Study*, RAND Corporation, Santa Monica, CA, P-4070, Apr 1969.

Oliver, S. A., J. Ausink, T. Manacapilli, J. Drew, S. Naylor, and C. Boone, *An Analysis of the Cost and Valuation of Aircraft Maintenance Personnel*, Air Force Logistics Management Agency, LM200107900, Jul 2002.

Perrin, B. M., D. S. Vaughan, J. L. Mitchell, R. M. Yadrick, and W. B. Bennett, *Methods for Clustering Occupational Tasks to Support Training Decision-Making*, Armstrong Laboratory, Brooks AFB, TX, AL/HR-TP-1996-0013, DTIC AD-A311 368, Jun 1996.

Robbert, A., J. Ausink, J. Ball, T. Manacapilli, G. Vernez, and N. Weaver, "Determining Course Content for Air Force Enlisted Initial Skill Training: How Can the System Be Made More Responsive?" RAND Corporation, Santa Monica, CA, internal draft, Jul 2001.

Samaniego, M. G., "Preliminary Plans for a Student Flow Model," RAND Corporation, Santa Monica, CA, working note, Jun 1972.

Selman, D. W., "Technical Training Systems," 2AF, briefing slides, 30 Jan 2002.

Slater, D., "The Whole . . . Is More Than Its Parts," *CIO Magazine*, http://www.cio.com/archive/051500/vision_content.html, 15 May 2000.

Sprague, Ralph H., Jr., and Eric D. Carlson, *Building Effective Decision Support Systems*, Prentice-Hall, Englewood Cliffs, NJ, 1982.

Weicher, M., W. W. Chu, W. C. Lin, V. Le, and D. Yu, *Business Process Reengineering Analysis and Recommendations, Communications Horizon*, http://www.netlib.com/bpr1.htm, Dec 1995.